Lecture Notes in Mathematics

Edited by A. Dold and B. Eckmann

1239

Paul Vojta

T0239438

Diophantine Approximations and Value Distribution Theory

Springer-Verlag

Berlin Heidelberg New York London Paris Tokyo

Author

Paul Vojta
Department of Mathematics, Yale University
New Haven, CT 06529, USA

Mathematics Subject Classification (1980): 11J68, 30D35

ISBN 3-540-17551-2 Springer-Verlag Berlin Heidelberg New York
ISBN 0-387-17551-2 Springer-Verlag New York Berlin Heidelberg

© Springer-Verlag Berlin Heidelberg 1987
Printed in Germany

Printing and binding: Druckhaus Beltz, Hemsbach/Bergstr.
2146/3140-543210

Introduction

Finding all solutions of a given system of diophantine equations has been shown to be an unsolvable problem, in general. More tractable, although still difficult, is the problem of determining whether the system has a finite number of solutions over every ring of integers of every number number field, possibly localized at a finite number of places. Or, one might ask the same question about k-rational solutions of the system.

The answer to both questions is known if the system of equations defines a curve, in the sense of algebraic geometry. Indeed, if the genus of the projective closure of the curve is zero, then there are always an infinite number of k-rational points for sufficiently large k, and similarly for integral points if there are at most two points at infinity. If there are three or more points at infinity, however, then finiteness always holds. If the genus is equal to one, then over a sufficiently large number field the curve is an elliptic curve with an infinite number of rational points, but finiteness always holds for integral points. Finally, if the genus is greater than one, then finiteness holds for rational as well as integral points.

In each case the answer depended only on algebraic-geometric invariants of the curve, and in fact, the "function field case" of the above questions has provided much insight.

More classically, though, the above invariants are also invariants of the associated Riemann surfaces, and the above answers have close parallels in the theory of holomorphic maps to Riemann surfaces. Indeed, there exists a nonconstant holomorphic map from $\mathbf{C}$ to a given compact Riemann surface of genus g if and only if $g \leq 1$ (Picard's theorem), and in the non-compact case, a Riemann surface of genus g missing $s > 0$ points admits such a map if and only if $g = 0$ and $s < 3$. Thus a curve has an infinite set of rational (resp. integral) points if and only if the associated compact (resp. non-compact) Riemann surface is the image of at least <u>one</u> non-constant holomorphic map.

These statements on holomorphic mappings can be proved using hyperbolic geometry; in fact Lang ([L 5], [L 8], see also §4.3) has posed a number of conjectures as higher dimensional equivalents of the above. For example:

CONJECTURE. *An algebraic variety V has the property that $V(k)$ is finite for all number fields k if, and only if, the associated complex space is Kobayashi hyperbolic.*

In this work we present a quantitative version of the above conjecture (the "Main Conjecture"). Of the above finiteness statements, those dealing with integral points can be proved using Roth's theorem, a statement in diophantine approximations. Moreover, the proofs give estimates on the minimum growth of the denominators of rational solutions. On the analytic side, the theorems dealing with the nonexistence of holomorphic functions all can be proved using a theory known as Nevanlinna theory; in the case of noncompact Riemann surfaces V, it provides asymptotic estimates for how many times a holomorphic map to $\overline{V}$ must meet $\overline{V} \setminus V$. Up to a point, the proofs of either set of theorems and the resulting estimates are very similar. This leads us to translate versions of Nevanlinna theory for higher dimensional varieties into quantitative versions of the above-mentioned conjectures of Lang. It should be noted that C. Osgood [Os 1] has also, previously, noted a Nevanlinna-Roth connection.

We also obtain a conjecture on all algebraic points of a variety (the "General Conjecture"); it is similar to the Main Conjecture, with an added term for the discriminant of the number field over which the point is defined. For example, Roth's theorem extends to a conjecture that, given $\epsilon > 0$ and points $\alpha_v \in k$ for finitely many places $v \in S$ of a number field k, then there exists a constant $c = c(k, S, \epsilon, \alpha_v)$ such that,

$$\prod_{v \in S} \prod_{\substack{w \in M_{k(x)} \\ w|v}} \min(1, \|x - \alpha_v\|_w) > \frac{c^{[k(x):k]}}{H(x)^{2+\epsilon} D_{k(x)/\mathbf{Q}}}$$

holds for almost all $x \in \mathbf{Q}$. Or, if C is a curve of genus ≥ 2 with canonical divisor K, then we conjecture that,

$$h_K(P) < \frac{1+\epsilon}{[k(P):\mathbf{Q}]} \log D_{k(P)/\mathbf{Q}} + O(1).$$

This special case of the General conjecture already implies a number of other conjectures–see Section 5.5.

Finally, in the last chapter we compare the proofs of theorems of Ahlfors and Schmidt, which give bounds on approximation to hyperplanes in projective space. The proofs are very similar; in particular, the use of differentiation in Ahlfors' proof is similar to the use of successive minima in Schmidt's.

Acknowledgements

This Lecture Note contains results obtained during three years spent at Yale University. It also contains some sections from the author's thesis, written at Harvard University. This note was completed at the Mathematical Sciences Research Institute in Berkeley.

I would like to thank Barry Mazur for much patience and encouragement, as well as guidance and ideas during the early stages of this work. It is also my pleasure to thank Serge Lang for much attention and many helpful suggestions during the later years. Without his energetic assistance, this Lecture Note might have languished in perpetual revision. I also thank Joe Silverman for many helpful comments. Finally, I thank Faye Yeager for asssistance with preparing the final manuscript.

This work was partially supported by an NSF postdoctoral fellowship and another NSF grant. I have also prepared parts of this book while staying briefly at the Institut Henri Poincaré in Paris and the Max-Planck Institut in Bonn.

Typeset by $\mathcal{A}_{\mathcal{M}}\mathcal{S}$-TEX.

Notations

C, **R**, **Q** The fields of complex, real, and rational numbers, respectively.

Z The ring of rational integers.

k Usually a number field. When discussing the function field case, this will also denote the field of constants.

M_k The set of all (inequivalent) absolute values of the global field k.

S_∞ The set of archimedean absolute values of k.

S A finite set of absolute values containing S_∞. Except for in Section 5.7, this does not necessarily contain the places of bad reduction.

v An element of M_k; if it is non-archimedean, the associated prime ideal is denoted $\wp$.

e The index of ramification of the non-archimedean place v.

f The degree of the residue class field extension of the local field k_v over $\mathbf{Q}_p$.

$\|\cdot\|_v$ $|\cdot|$, if $k_v = \mathbf{R}$; $|\cdot|^2$, if $k_v = \mathbf{C}$, or $p^{-f \operatorname{ord}_\wp(\cdot)}$, if v is non-archimedean.

O_k The ring of integers of the number field k.

$O_{k,S}$ The ring O_k with primes in S inverted.

λ_v Weil function, normalized as in §1.3.

$|D|$ The support of a divisor D; i. e. the divisor obtained by changing all nonzero multiplicities to one.

$|S|$ The number of elements of the set S.

$D \geq E$ When D and E are divisors, this means that $D - E$ is effective.

Also, the notations of algebraic geometry are those of [**H**], except that varieties are often taken as defined over number fields.

Other notations appear in the index.

Contents

Chapter 1. Heights and Integral Points   1

  §1. Absolute values and the product formula 1
  §2. Height functions 2
  §3. Weil functions 7
  §4. Integral points 10

Chapter 2. Diophantine Approximations   16

  §1. Roth's theorem and its generalizations 16
  §2. A reformulation of Schmidt's theorem 17
  §3. Linear equations in units 19
  §4. A theorem on integral points 21
  §5. Arithmetic distance functions and Schmidt's theorem . . . 24
  §6. Exceptional hyperplanes 27

Chapter 3. A Correspondence with Nevanlinna Theory   30

  §1. Introduction to Nevanlinna theory 31
  §2. The correspondence in the case of one variable 33
  §3. Defects 36
  §4. The higher dimensional case 38
  §5. Some examples 43

Chapter 4. Consequences of the Main Conjecture   46

  §1. Degeneracy of integral points 46
  §2. Integral points on abelian varieties 47
  §3. Lang's conjectures on curvature and hyperbolicity 48
  §4. Hall's conjecture 51

Chapter 5. The Ramification Term   57

  §1. A generalized Chevalley-Weil theorem 57
  §2. The ramification term 61
  §3. The $\mathbf{P}^1$ case of the General Conjecture 65
  §4. Other remarks on the General Conjecture 68
  §5. Examples: Fermat, abc, Hall, and Hall-Lang-Stark 70
  §6. The (split) function field case 77
  §7. The ramification term in the Conjecture using (1,1) forms . 80
  Appendix ABC. The Masser-Oesterlé "abc" conjecture 84

Chapter 6. Approximation to Hyperplanes 89

§1. Successive minima 90

§2. Davenport's lemma 96

§3. Multilinear algebra 99

§4. The start of the algebraic proof 104

§5. A sketch of the analytic proof 112

§6. The remainder of the algebraic proof 118

§7. Conclusion 122

Bibliography 124

Index 130

Chapter 1
Heights and Integral Points

§1. Absolute Values and the Product Formula

Let k be a number field. An absolute value on k is a real-valued function $|\cdot|$ on k satisfying the following properties:

(1). $|x| \geq 0$, and $|x| = 0$ if and only if $x = 0$;

(2). $|xy| = |x||y|$;

(3). $|x + y| \leq 2 \max(|x|, |y|)$.

Absolute values are also sometimes called places or primes. An embedding $\phi : k \to \mathbf{C}$ causes the absolute value on $\mathbf{C}$ to induce an absolute value on k; such absolute values are called infinite, or archimedean, primes.

Finite, or non-archimedean, places come from a prime ideal $\wp$ of $\mathcal{O}_k$, the ring of integers of k. Such places are defined by,

$$|x|_\wp = c^{-\operatorname{ord}_\wp(x)}$$

where $c > 1$ is a real constant and $\operatorname{ord}_\wp(x)$ is the power of $\wp$ appearing in the prime factorization of the principal ideal (x). For the field $\mathbf{Q}$, it is customary to take $c = p$; for finite extensions of $\mathbf{Q}$, two options are possible, denoted $|\cdot|$ and $\|\cdot\|$. They are defined such that, if p is the rational prime associated to $\wp$,

$$(1.1.1) \qquad |p| = p^{-1} \quad \text{and} \quad \|p\| = p^{-[k_\wp : \mathbf{Q}_p]}.$$

Instead of condition (3) above, non-archimedean places are characterized by the fact that they satisfy the stronger inequality,

(3'). $|x + y| \leq \max(|x|, |y|)$.

If v is an absolute value, it induces on k the structure of a metric space, and thus a topology. Two absolute values are said to be equivalent if they induce the same topology; in that case one is a power of the other. The completion of k relative to the topology induced by v is a field which is denoted k_v. If v is archimedean, then $k_v = \mathbf{R}$ or $\mathbf{C}$; if it is non-archimedean, then k_v is a finite extension of $\mathbf{Q}_p$, which explains the notation $[k_\wp : \mathbf{Q}_p]$ in (1.1.1).

For all absolute values v, we define the normed absolute value $\|x\|_v$ for $x \in k$ by,

$$(1.1.2) \qquad \|x\|_v = \begin{cases} |x|, & \text{if } k_v = \mathbf{R}; \\ |x|^2, & \text{if } k_v = \mathbf{C}; \\ \text{See } (1.1.1), & \text{if } v \text{ is non-archimedean.} \end{cases}$$

Under these conventions there exists a set M_k of inequivalent primes of k satisfying a product formula,

$$(1.1.3) \qquad \prod_{v \in M_k} \|x\|_v = 1 \qquad \text{for all } x \in k^*.$$

This set M_k consists of one v for each prime ideal of $\mathcal{O}_k$ and one prime for each real or conjugate pair of complex embeddings. This product formula is often stated with multiplicities, but by the choices $(1.1.2)$, these multiplicities are unnecessary.

The subset of archimedean places $v \in M_k$ is denoted S_∞. If F is a finite extension of k, then places $w \in M_F$ restrict to places $v \in M_k$ (up to a fixed power). Denote this by writing $w \mid v$. The product formulas of F and k correspond term by term in the sense that,

$$(1.1.4) \qquad \prod_{\substack{w \mid v \\ w \in M_F}} \|x\|_v = \|x\|_v^{[F:k]}.$$

The product formula $(1.1.3)$ can be used to axiomatically describe a global field; i. e. a number field or the function field of a variety. In the sequel, however, function fields will play only a minor role and therefore we consider primarily number fields.

§2. Height Functions

The so-called naïve height, or the Weil height, is defined on $\mathbf{P}^n(k)$ by choosing homogeneous coordinates $(x_0, \ldots, x_n)$ for P and letting

$$(1.2.1) \qquad H(P) = \prod_{v \in M_k} \max(\|x_0\|_v, \ldots, \|x_n\|_v).$$

By the product formula, $H(P)$ is independent of the choice of homogeneous coordinates for P.

Note, however, that $H(P)$ does depend on the number field in question; in fact, if F is a finite extension of k, then by (1.1.4),

$$H_F(P) = H_k(P)^{[F:k]}.$$

Thus, it is called the <u>relative height</u>. It is possible to make a definition of height independent of the number field; it is convenient to also take the logarithm, defining,

$$h(P) = \frac{1}{[k:\mathbf{Q}]} \log H(P),$$

where the height on the right is defined over k. This is called the (<u>absolute</u>) <u>logarithmic height</u>. By contrast, the height $H(P)$ is called the (<u>relative</u>) <u>multiplicative height</u>.

DEFINITION 1.2.2. *Let f and g be two non-negative functions. If $f < cg$ for some positive constant c, then we write $f \ll g$ or $g \gg f$. The notation $f \gg\ll g$ means that both $f \ll g$ and $f \gg g$ hold.*

DEFINITION 1.2.3. *Two heights H_1 and H_2 (resp. logarithmic heights h_1 and h_2) are called <u>equivalent</u> if $H_1 \gg\ll H_2$ (resp. $h_1 = h_2 + O(1)$).*

The remainder of this section will use only the logarithmic height; similar comments obviously hold for the multiplicative height.

We now wish to define a notion of height on a projective variety V defined over k. (For the remainder of this section, assume that all varieties, divisors, functions, etc. on V are defined over k.) To define the height, we use a projective embedding of V, together with (1.2.1). Therefore the definition of the height depends on choices of a very ample divisor D and functions in the linear system $\mathcal{L}(D)$. Up to equivalence, however, the height depends only on D and furthermore,

LEMMA 1.2.4. *If D and D' are two very ample divisors on V, then*

$$h_{D+D'} = h_D + h_{D'} + O(1).$$

PROOF: See [**L 7**]. $\qquad\qquad\qquad\qquad\qquad\qquad\qquad\qquad\qquad\qquad\qquad\square$

Now, given any divisor D, we can write $D = E - E'$ where E and E' are very ample, and define

$$h_D = h_E - h_{E'}.$$

This definition depends on the choices of E and E', but by Lemma 1.2.4 h_D is well defined up to equivalence. Furthermore, it then follows that (1.2.4) holds for arbitrary divisors D and D'.

In order to summarize the properties of heights it will be necessary to define some terminology, starting with a definition due to Iitaka [Ii, Ch. 10, Theorem 10.2]:

DEFINITION 1.2.5.

 (a). *Let D be a divisor on a nonsingular variety V. The <u>dimension of</u> <u>D</u> is the integer $d = \dim D$ such that*

$$\ell(nD) \gg\!\!\ll n^d.$$

 for n sufficiently divisible. If $\mathcal{L}(nD)$ is always empty, then let $d = -\infty$.

 (b). *D is <u>almost ample</u> if $\dim D = \dim V$.*

Thus, for example, a variety of general type is one for which the canonical divisor is almost ample.

LEMMA 1.2.6. *The dimension has the following properties:*

 (a). *If D, E are divisors and if D is effective, then $\dim(D+E) \geq \dim E$; in particular, $\dim D \geq 0$.*

 (b). *$\dim D \leq \dim V$.*

 (c). *If D is ample then $\dim D = \dim V$.*

 (d). *The dimension depends only on the linear equivalence class of D; in particular it can be defined for a line bundle.*

 (e). *If $f: V \to W$ is a proper surjective morphism of nonsingular complete varieties and $\mathcal{L}$ is a line bundle on W, then $\dim f^*\mathcal{L} = \dim \mathcal{L}$.*

PROOF: (a)–(d) are trivial; for (e) see [Ii, Theorem 10.5.]   □

EXAMPLE. Let V be a surface and let E be a curve on V with negative self intersection. Let D be an ample divisor on V. Then $D+mE$ is almost ample for all m, but it is not ample if $(E^2)m > (E.D)$. Thus not all almost ample divisors are ample.

PROPOSITION 1.2.7 (KODAIRA, [K-O, APPENDIX]). *Let D be a divisor on a projective variety V. Then D is almost ample if and only if*

nD can be written as a sum of an ample divisor and an effective divisor, for some sufficiently large integer n.

PROOF: If D can be written as a sum of an ample divisor and an effective divisor, then it is almost ample, by Lemma 1.2.6. Conversely, let A be any ample divisor, and assume that it is irreducible and smooth. Then we have an exact sequence,

$$0 \to \mathcal{O}(-A) \to \mathcal{O}_V \to \mathcal{O}_A \to 0$$

where $\mathcal{O}(-A)$ is the invertible sheaf associated to the divisor $-A$. Tensoring with $\mathcal{O}(nD)$ gives a long exact sequence in cohomology,

$$0 \to H^0(V, \mathcal{O}(nD - A)) \to H^0(V, \mathcal{O}(nD)) \to H^0(A, \mathcal{O}(nD)|_A) \to \ldots$$

Counting dimensions, the second term grows like $n^{\dim V}$, whereas the third grows only at most as fast as $n^{\dim V - 1}$. Thus, $h^0(V, \mathcal{O}(nD - A)) > 0$ for some n sufficiently large, as was to be shown. □

DEFINITION 1.2.8. Let h_1 and h_2 be two logarithmic height functions on V, and let D be any ample divisor. If for any $\epsilon > 0$ there exists a constant $c = c(\epsilon)$ such that,

$$|h_1 - h_2| < \epsilon h_D + c,$$

then we say that h_1 and h_2 are quasi-equivalent.

This notion is independent of the choice of D, by (f), below. We can now summarize the properties of heights:

PROPOSITION 1.2.9.

(a). For all divisors D, D' on V,

$$h_{D+D'} = h_D + h_{D'} + O(1).$$

(b). If D and D' are linearly equivalent, then

$$h_D = h_{D'} + O(1).$$

(c). If $f: V \to W$ is an algebraic map defined over k, and D is a divisor on W, then

$$h_{V, f \cdot D} = h_{W, D} \circ f + O(1).$$

(d). If D and D' are two numerically equivalent divisors, then h_D and $h_{D'}$ are quasi-equivalent.

(e). If D is effective, then $h_D \geq O(1)$ outside of the base locus of D.

(f). If D is ample then h_D is the largest possible height function, up to a constant; i. e. for any other divisor E,

$$h_D \gg h_E + O(1).$$

(g). (Northcott) if D is ample then there are only finitely many k-rational points of V for which h_D is below a given bound.

(h). If D is almost ample, then (f) and (g) hold outside of a Zariski-closed subset of V.

PROOF: (a) has been discussed. For (b), (c), (e), and (f) see [**L 7**]; Ch. 4; 2.1, 5.1, 5.2, and 5.4, respectively. For (g), see [**L 7**, Ch. 3, Theorem 2.6.]

(d). By (a) it will suffice to show that if D is numerically equivalent to zero and L is an ample divisor, then for all $\epsilon > 0$ there exists a constant c such that,

$$|h_D| \leq \epsilon h_L + c.$$

But, for any positive integer n, $L - nD$ is ample. Indeed, by the Nakai-Moishezon criterion [**H**, A.5.1], ampleness depends only on the numerical equivalence class of V; since $L - nD$ is numerically equivalent to L, $L - nD$ is ample. Thus by (b) and (e),

$$h_{L-nD} \geq O(1);$$

$$h_D \leq \frac{1}{n} h_L + c.$$

By the same argument applied to $-D$, $-h_D \leq \frac{1}{n} h_L + c$. This concludes the proof of (d).

(h). Follows from (f), (e), (c), and Proposition 1.2.7. □

REMARK. At first glance, (d) may appear more general than the usual property of heights being quasi-equivalent if their divisors are algebraically equivalent [**L 2**]. However, by [**Mat**], numerical equivalence is the same as algebraic equivalence up to torsion, so the results are actually identical.

By (h), an almost ample divisor will give a height function which is "largest," even on a complete variety which is not projective.

§3. Weil Functions

Let k, M_k, and S_∞ be as above; let D be a divisor on a nonsingular variety V. Extend $\|\cdot\|_v$ to an absolute value on the algebraic closure $\overline{k}_v$. Then a <u>local Weil function</u> for D relative to v is a function $\lambda_{D,v}: V(\overline{k}_v) \setminus |D| \to \mathbf{R}$ such that if D is represented locally by (f) on an open set U, then

$$\lambda_{D,v}(P) = -\frac{1}{[k:\mathbf{Q}]} \log \|f(P)\|_v + \alpha(P)$$

where $\alpha(P)$ is a continuous function on $U(\overline{k}_v)$. We sometimes think of $\lambda_{D,v}$ as a function of $V(k) \setminus |D|$ or $V(\overline{k}) \setminus |D|$ by (implicitly) choosing an embedding $\overline{k} \to \overline{k}_v$.

This is very similar to the definition for a metric. Indeed, let (U_i, f_i) be a Cartier divisor representing D. Let the associated line bundle $[D]$ be defined by trivializations $\phi_i: [D]|_{U_i} \to U_i \times \mathbf{C}$ such that

$$\phi_j \circ \phi_i^{-1}: (u, x) \mapsto (u, f_j(u)/f_i(u) \cdot x)$$

on $[D]|_{U_i \cap U_j}$. Then a metric on the line bundle $[D]$ is a set of C^∞ functions,

$$\rho_i: U_i \to \mathbf{R}$$

satisfying,

(1.3.1) $$\rho_i/\rho_j = |f_i/f_j|^2 \qquad \text{on } U_i \cap U_j.$$

The functions f_i give a section s of the line bundle $[D]$ whose divisor is D. Then, given $P \in V$, we have

$$|s(P)| = \frac{|f_i(P)|}{\rho_i(P)}$$

so that

(1.3.2)
$$\begin{aligned}
\lambda_{D,v}(P) &= -\frac{1}{[k:\mathbf{Q}]} \log \|s(P)\|_v \\
&= -\frac{1}{[k:\mathbf{Q}]} \log \|f_i(P)\|_v + \frac{1}{[k:\mathbf{Q}]} \log \rho_{i,v}(P)
\end{aligned}$$

can be taken to be a local Weil function for D at v.

A <u>global Weil function</u> for D over k is a collection $\{\lambda_D\}$ of local Weil functions, as v ranges through M_k, subject to an additional continuity constraint. This constraint is fairly hard to state (see [L 7, Ch. 10, §1, §2]); however, for our purposes it suffices to know that changing a finite number of local Weil functions will not affect the continuity condition. Note also that we normalize the Weil functions differently here.

LEMMA 1.3.3. *Weil functions satisfy the following properties:*

(a). *If λ_D and $\lambda_{D'}$ are (local or global) Weil functions for D and D', then $\lambda_D + \lambda_{D'}$ is a Weil function for $D + D'$ and $-\lambda_D$ is a Weil function for $-D$.*

(b). *If D is an effective divisor then $\lambda_{D,v}(P) \geq c_v$ for some constant c_v; furthermore $c_v = 0$ for almost all v.*

(c). *$-\frac{1}{[k:\mathbf{Q}]} \log \|f\|_v$ is a global Weil function for the principal divisor (f).*

(d). *Let $\phi: V \to W$ be a morphism of nonsingular varieties and let D be a divisor on W not containing the image of ϕ. If λ_D is a Weil function for D on W then $\lambda_D \circ f$ is a Weil function for f^*D on V.*

(e). *If E is a finite extension of the number field k and λ_D is a Weil function for D over k then*

$$\lambda_{D,w}(P) = \frac{[E_w : k_v]}{[E : k]} \lambda_{D,v}(P)$$

is a Weil function for D over E. Thus, if $P \in V(k)$, then

$$\lambda_{D,v}(P) = \sum_{w|v} \lambda_{D,w}(P) + O(1).$$

(f). *Let $D_1, \ldots, D_n$ and D be divisors on V such that $D_i - D$ are effective divisors with no geometric point in common. Then*

$$\inf_i \lambda_{D_i}$$

is a Weil function for D.

(g). *The height can be expressed in terms of Weil functions as,*

(1.3.4) $$h_D(P) = \sum_{v \in M_k} \lambda_{D,v}(P) + O(1),$$

for all $P \in V(k) \setminus |D|$.

PROOF: See [**L 7**]. □

For example, (c) and (f) imply that,

(1.3.5) $$\lambda_{D,v}(P) = \frac{1}{[k : \mathbf{Q}]} \log \max(1, \|x_1/x_0\|_v, \ldots, \|x_n/x_0\|_v)$$

is a Weil function for the hyperplane $(x_0 = 0)$ at infinity on $\mathbf{P}^n$.

Hermitian metrics on line bundles satisfy a similar list of properties. In later chapters, most of the hermitian metrics encountered will be of the type in which,

$$\rho_i = \frac{|\alpha_1|^2 + |\alpha_2|^2 + \cdots + |\alpha_n|^2}{(|\beta_1|^2 + \cdots + |\beta_m|^2)(\log \mu/(|\gamma_1|^2 + \cdots + |\gamma_r|^2))^2},$$

where the α's and β's are in $O(U_i)$, $\mu \geq 1$, and the relation (1.3.1), $\rho_i/\rho_j = |f_i/f_j|^2$, can be deduced formally from the above expressions for ρ_i and ρ_j, using only the rule $|\alpha\beta|^2 = |\alpha|^2|\beta|^2$. These metrics have mild singularities at the poles and common zeroes of the γ's.

It is possible to translate such metrics into Weil functions in a natural way. Let D be a divisor and let $s = \{f_i\}$ be a section of the line bundle $[D]$ with divisor $(s) = D$. Then in place of (1.3.2) we can define,

$$\lambda_{D,v}(P) = -\frac{1}{[k:\mathbf{Q}]} \log \|f_i\|_v$$

$$+ \lim_{E} \sum_{\substack{w \in M_E \\ w|v}} \frac{1}{[E:\mathbf{Q}]} \log \frac{\|\alpha_1\|^2 + \|\alpha_2\|^2 + \cdots + \|\alpha_n\|^2}{(\|\beta_1\|^2 + \cdots + \|\beta_m\|^2)(\log \mu/(\|\gamma_1\|^2 + \cdots + \|\gamma_r\|^2))^2}$$

The limit over E is with respect to the directed system of number fields, ordered by inclusion. For example, the Fubini-Study metric on $\mathbf{P}^n$ is given by,

$$\rho_i = \frac{\|x_0\|_w + \cdots + \|x_n\|_w}{\|x_i\|_w}, \qquad 0 \leq i \leq n.$$

For archimedean places, the limit stabilizes and gives $(1/2)\log|s(P)|^2$ as above; for non-archimedean places, the limit reduces to a computation of,

$$\lim_{ef \to \infty} \sqrt[ef]{a_0^{ef} + \cdots + a_n^{ef}} = \max(a_0, \ldots, a_n)$$

for nonnegative real numbers $a_0, \ldots, a_n$. The associated Weil function for a hyperplane on $\mathbf{P}^n$ is then,

$$\lambda_{D,v}(P) = \begin{cases} -\frac{1}{2[k:\mathbf{Q}]} \log(1 + |x_1/x_0|^2 + \cdots + |x_n/x_0|^2), & v \mid \infty; \\ \frac{1}{[k:\mathbf{Q}]} \log \max(1, \|x_1/x_0\|_v, \ldots, \|x_n/x_0\|_v), & v \nmid \infty. \end{cases}$$

The computation is similar in the general case, noting that the logarithm drops out for non-archimedean v.

The limit might seem unnatural, but one could think of it as a Riemann integral.

§4. Integral Points

Fix a number field k. As before, assume from now on that all varieties, maps, etc. are defined over k. Let S be a finite set of places containing the set S_∞ of all archimedean places. Let $\mathcal{O}_S$ be the ring of S-integers of k; i. e. the set of all $x \in k$ such that $\|x\|_v \leq 1$ for all $v \notin S$. It is clear, then, that a point $P \in \mathbf{A}^n(k)$ should be called an integral point if and only if all its coordinates are S-integers, and that an algebraic point $P \in \mathbf{A}^n(\overline{k})$ should be integral if its coordinates lie in the integral closure of $\mathcal{O}_{k,S}$ in $\overline{k}$. Similarly, an affine variety $W \subseteq \mathbf{A}^n$ defined over k inherits a notion of integral point from the definition for $\mathbf{A}^n$.

It is possible to formulate this definition more intrinsically, though. Let V be a projective variety, let D be a very ample effective divisor on V, and let $1 = x_0, x_1, \ldots, x_n$ be a basis for $\mathcal{L}(D)$. Then $P \mapsto (x_1(P), \ldots, x_n(P))$ defines an embedding of $V - D$ into $\mathbf{A}^n$; therefore we say P is an integral point if $x_i(p) \in \mathcal{O}_S$ (or its integral closure in $\overline{k}$) for all i. Although this notion of integral point actually depends on k, S, D, and $\{x_i\}$, we shorten the language and say that P is a D-integral point or that P is an integral point relative to D.

One must be careful, though: any point P on $V(k) \setminus D$ can be a D-integral point for some basis of $\mathcal{L}(D)$. Indeed, let $\{x_i\}$ be any basis and let $b \in \mathcal{O}_k$ be a number which clears the denominators of $x_i(P), 1 \leq i \leq n$. Then $\{1, bx_1, \ldots, bx_n\}$ is a basis of $\mathcal{L}(D)$ for which P is integral. Therefore, we let integrality be a property of the <u>set</u> of points. This is a natural concept in light of the following lemma.

LEMMA 1.4.1. *Let D be a very ample effective divisor on V. Let $\mathcal{R}$ be a subset of $V(k) \setminus |D|$. Then the following are equivalent.*

(a). *$\mathcal{R}$ is a set of D-integral points on V;*
(b). *There exists a global Weil function $\lambda_{D,v}$ and constants c_v for each $v \in M_k \setminus S$, such that (i) almost all $c_v = 0$, and (ii) for all $P \in \mathcal{R}$, all $v \in M_k \setminus S$, and all embeddings of $\overline{k}$ in $\overline{k}_v$,*

$$\lambda_{D,v}(P) \leq c_v.$$

PROOF: Let $1 = x_0, x_1, \ldots, x_n$ be a basis for $\mathcal{L}(D)$ such that $x_i(P)$ is integral over $\mathcal{O}_{k,S}$ for all $1 \leq i \leq n$ and all $P \in \mathcal{R}$. Then, letting

$$\lambda_{D,v}(P) = \frac{1}{[k:\mathbf{Q}]} \log \max(\|x_0(P)\|_v, \ldots, \|x_n(P)\|_v)$$

we see that $\lambda_{D,v}(P) \leq 0$. Conversely, let $\lambda_{D,v}$ be as above and assume it is bounded by c_v. Clearing denominators in the $x_i(P)$, we obtain $\|x_i(P)\|_v \leq 1$ for all $v \notin S$. Then $\mathcal{R}$ is a set of integral points relative to the new $x_1, \ldots, x_n$. $\qquad\square$

COROLLARY 1.4.2. *The notion of D-integrality is independent of the multiplicities of the components of D.*

PROOF: Immediate, using the above lemma and Lemma 1.3.3 (a) and (b).
$\qquad\square$

Thus the notion of integrality depends only on the set $V \setminus |D|$. This lemma motivates a more general definition of integrality:

DEFINITION 1.4.3. *Let D be an effective divisor on V and let $\mathcal{R}$ be a subset of $V(\overline{k}) \setminus |D|$. Then $\mathcal{R}$ is an __(S,D)-integralizable__ set of points if there exists a global Weil function satisfying condition (b) of Lemma 1.4.1.*

REMARK 1.4.4.

- (a). Often S will be fixed and we will refer to a D-integralizable set of points.
- (b). By abuse of language, we will often use the wording, "set of integral points" instead of "integralizable set of points."
- (c). Condition (b) of Lemma 1.4.1 holds for some choice of Weil function if and only if it holds for all choices of Weil functions (with suitable choices of constants c_v.)

This definition also makes sense if D is <u>any</u> effective divisor, even if it is not ample. In particular, if $D = 0$ then "D-integral" means "rational." This will allow us to discuss integral points on V if $V \setminus D$ is neither projective nor affine. It is useful to discuss such varieties because they occur as moduli spaces. Or, suppose a system of diophantine equations gives an affine variety which is singular. One could either work with singular varieties, or resolve the singularities and work with non-affine varieties. In the sequel we will choose the latter option.

As easy consequences of the properties of Weil functions, we find:

LEMMA 1.4.5. *Let S be a finite set of places of k containing the archimedean places, let E be a finite extension field of k, and let T be the set of places of E lying over places in S. Then $\mathcal{R} \in V(\overline{k})$ is a set of (S,D)-integral points if and only if it is a set of (T,D)-integral points.*

LEMMA 1.4.6. *Let R be a D-integral set of points on V and let f be a rational function with no poles outside of D. Then there is some constant $b \in k$ such that $bf(P)$ is integral for all $P \in R$.*

Integral points can also be discussed on schemes over Spec $O_{k,S}$. It would be instructive here to relate the two concepts. Let X be a scheme of finite type over $Y = $ Spec O_S. Then a scheme-theoretic integral point of X is a section $s: Y \to X$ of the map $X \to Y$, and an algebraic integral point is an integral point on some finite base change of X. Assume that $X \times_Y$ Spec k is nonsingular, and let V be a nonsingular complete variety such that $V \setminus D = X \times_Y$ Spec k, where D is a Zariski-closed subset which we may assume to be a divisor (by blowing it up to codimension 1). The following lemma relates integrality in the scheme sense with the notion discussed above.

PROPOSITION 1.4.7. *Let X, Y, V, and D be as in the above paragraph. Then, after enlarging S, the following two statements hold.*

(a). *Let R be the set of algebraic integral points of X (in the sense of schemes over Spec O_S). Then the corresponding set R_k of points on $V(\overline{k})$ is a (D, S)-integral set of points (in the sense of Definition 1.4.3).*

(b). *Let R be a (D, S)-integral subset of $V(\overline{k})$. Then there exists a scheme X' over Y such that all points $P \in R$ correspond to algebraic S-integral points of X'.*

PROOF: By [**Na**], X can be embedded in a scheme $\overline{X}$, proper over Y. Let $D = \overline{X} \setminus X$; by blowing up, we may assume D is a divisor. Let $\overline{X}_k = \overline{X} \times_Y$ Spec k. We identify D with the corresponding divisor on $\overline{X}_k$ and points $P \in R$ with the corresponding points in $\overline{X}_k(\overline{k})$. We enlarge S (*i. e.* shrink Y) so that $\overline{X}$ is regular and all fibers of D reduce to normal crossings divisors on the fibers of $\overline{X}$. Then, as in [**L 7**, Ch. 11, Thm. 5.1], we can define a Weil function by using the intersection pairing. Indeed, let v be a place of Y and let q_v be the number of elements in the residue field of O_v. Let $\langle -.- \rangle$ denote the usual (integer-valued) intersection number, multiplied by $\log q_v$. If P is rational over Y, then let

$$\lambda_{D,v}(P) = \frac{1}{[k : \mathbf{Q}]} \langle P.D \rangle_v$$

$$= \frac{1}{[k : \mathbf{Q}]} (f_v \log p) \, \mathrm{length}_{O_v}(O_v/(f(P))).$$

Here f is a local equation for D; the second line is to be taken as the definition of the intersection pairing. If P is an algebraic point, defined over a base Y' finite over Y, and w is a point of Y' lying over v (corresponding to a choice of embedding $E_w \to \overline{k}_v$), then let

$$\lambda_{D,v}(P) = \frac{1}{[E_w : k_v][k : \mathbf{Q}]} \langle P.D \rangle_w.$$

This notion is well-defined by the basic properties of intersection theory.

Now the proof of (a) is easy. Indeed, points of $\mathcal{R}$ do not meet D; therefore $\lambda_{D,v}(P) = 0$ for such points.

To prove (b), we may enlarge X if necessary so that no components of D lie in the fibers of the map to Y. Let $\lambda_{D,v}$ be the Weil function as above; by assumption and by Remark 1.4.4 (c), there exist constants c_v, almost all zero, such that $\lambda_{D,v}(P) \leq c_v$ for all points $P \in \mathcal{R}$ and all embeddings of $\overline{k}$ in $\overline{k}_v$. We obtain the desired scheme X' by repeated applications of the following lemma.

LEMMA 1.4.8. *Let* $\overline{X} \to Y$, D, *and* X *be as above, and let* D_0 *be a component of* D. *Let* $v \in Y$. *Let* D_v *denote the intersection of* D_0 *with the fiber over* v, *and let* $\pi : \overline{X}' \to \overline{X}$ *be the blowing-up of* $\overline{X}$ *with center* D_v. *Let* D', D'_0 *be the strict transforms of the divisors* D *and* D_0, *respectively, and let* E *be the exceptional divisor of* π. *Let* $\lambda'_{D,v}$ *denote the Weil function defined as above relative to the intersection pairing on* $\overline{X}'$. *Then:*

(a). *If* $w \neq v$, *then*

$$\lambda'_{D,w}(P) = \lambda_{D,w}(P);$$

(b). *for any component* D_1 *of* D,

$$\lambda'_{D_1,v}(P) \leq \lambda_{D_1,v}(P);$$

(c). *if* $\lambda_{D_0,v}(P) > 0$ *then*

$$\lambda'_{D_0,v}(P) \leq \lambda_{D_0,v}(P) - \frac{\log q_v}{[k : \mathbf{Q}]}.$$

PROOF: Part (a) follows from the fact that π is an isomorphism outside of D_v.

If D_1' denotes the strict transform of D_1, then $\pi^* D_1 = D_1' + rE$ for some integer $r \geq 0$. Then (b) follows from the fact that $\langle P.E \rangle \geq 0$ and functoriality and linearity of the intersection pairing.

Finally, for part (c) assume P' is rational over Y' and let $w \in Y'$ lie over $v \in Y$. Let e and f denote the ramification index at w and the degree of the residue field extension, respectively. Then $\pi^* D_0 = D_0' + rE$ pulls back to $D_0'' + er E''$ on $\overline{X} \times_Y Y'$. Since $\lambda_{D_0,v}(P) > 0$, P meets E'' and therefore

$$\langle P.er E'' \rangle \geq er \log q_w.$$

Since $q_w = q_v^f$, $[E_w : k_v] = ef$, and $r \geq 1$,

$$\lambda_{D_0,v}(P) - \lambda'_{D_0,v}(P) \geq \frac{1}{[k : \mathbf{Q}]} \log q_v. \qquad \square$$

When there is a theory of resolution of singularities (e. g. when V is a curve over a number field, or in the case of function fields of characteristic zero), then it is not necessary to enlarge S at all.

In the context of schemes, it is clear that under a map of Y-schemes $X \to X'$, integral points are taken to integral points. This can also be proved in our context, without referring to schemes. Indeed, let $f: V \setminus D \to W \setminus E$ be a morphism, which we may assume extends to a morphism defined on all of V. Since $f^* E \subseteq D$, we see again by Lemma 1.3.3 (d) that sets of integral points are taken to sets of integral points. Thus, on a formal level, infinite sets of D-integral points on V behave much like holomorphic maps to the complex manifold $V \setminus D \times \operatorname{Spec} \mathbf{C}$. This will be developed further in Chapter 3, but here we note one more instance of this principle.

THEOREM 1.4.9 (CHEVALLEY-WEIL). *Let* $\pi: W \setminus E \to V \setminus D$ *be an étale map and let* $\mathcal{R}$ *be a D-integral subset of $V(k)$. Then the discriminants of the field extensions $k(\pi^{-1}(P))/k$ are bounded for all $P \in \mathcal{R}$.*

PROOF: See [**L 7**, Theorem 8.1]. A more general proof is also given in Chapter 5. $\qquad \square$

THEOREM 1.4.10 (HERMITE-MINKOWSKI). *A field k has only finitely many extensions F of bounded degree and bounded discriminant.*

PROOF: See [**L 4**, Theorem 5, Ch. 5, §4]. $\qquad \square$

THEOREM 1.4.11. *Let k and $f: W \setminus E \to V \setminus D$ be as above, and let $\mathcal{R}$ be a set of (S, D)-integral points of V. Then there exists a number field F (depending on $\mathcal{R}$) such that $\pi^{-1}(\mathcal{R}) \subseteq W(F)$. Moreover, if T is the set of places of F lying over places of S, then $\pi^{-1}(\mathcal{R})$ is a set of (T, E)-integral points of W.*

PROOF: The first statement comes from the previous two theorems; the second, from the definitions and Lemma 1.3.3 (d). □

REMARK 1.4.12. This corresponds to the fact that a holomorphic map to a manifold may be lifted to a finite covering space by taking a finite cover of the domain so as to eliminate the obstruction in π_1 (the principle of monodromy for continuation of analytic functions).

The conclusions of the theorems on integral points on curves assert that all sets of integral points are finite. For higher dimensional varieties, one would hope to prove similar statements on finiteness of integral points. It is more natural, however, to try to find conditions implying that all sets of integral points are contained in a subvariety. Being a weaker assertion, it is easier to prove; also, it can lead to a finiteness statement by reducing the problem to several problems of lower dimension. Therefore, make the following definition.

DEFINITION 1.4.13. *A set of k-rational points of V is called <u>degenerate</u> if it is not dense in the Zariski topology.*

Chapter 3 will explain the choice of the word "degenerate."

Chapter 2
Diophantine Approximations

In this chapter we introduce Roth's theorem, one of the most commonly used theorems in diophantine approximations. We will be most concerned with the generalizations to approximations to hyperplanes in projective space, due to Schmidt and Schlickewei. These will be introduced in the first section of the chapter; the second section describes a reformulation which will be more useful in the sequel. In the third section, we will state and prove a result on linear equations in units, due to van der Poorten and Schlickewei.

In the fourth section, we use this result to prove a theorem on integral points on varieties relative to a divisor having a sufficiently large number of components. This number depends on the dimension of the variety, the rank of the Néron-Severi group, and the rank of the group of k-rational points of the Picard variety Pic^0. When the latter variety is trivial, the number of components required is independent of the field of definition of the variety. These conditions, as well as the proof of the theorem, bear a close resemblance to a theorem of Mark Green, dealing with holomorphic maps to complex varieties. It is this similarity that led to the conjecture of Chapter 3.

In Section 5, we introduce more general versions of Weil functions, called arithmetic distance functions, which allow Schmidt's theorem to be restated again in a form very similar to what will be used in Chapter 3. Finally, in the last section of the chapter, we use this mechanism to examine the exceptional hyperplanes mentioned in the statement of Schmidt's theorem.

§1. Roth's Theorem and its Generalizations

With the exception of the Mordell conjecture, Roth's theorem can be used to prove all known general results on integral points on curves. It is the following.

THEOREM 2.1.1. *Let k be a number field and S a finite set of places of k. Let $\epsilon > 0$ and $c > 0$ be real constants. For all $v \in S$, fix $a_v \in \overline{\mathbf{Q}}$. Let $H(x)$ denote the relative multiplicative height. Then the inequality,*

$$\prod_{v \in S} \min(1, \|x - a_v\|_v) < \frac{c}{H(x)^{2+\epsilon}},$$

holds for only finitely many $x \in k$.

This was generalized to higher dimensions by Schmidt ([**Schm 1**]) and in the non-archimedean case by Schlickewei ([**Schl 1**], [**Schl 2**], [**Schl 3**]). The final result is stated as Theorem 2.1 of [**Schl 3**]. With minor changes in notation, it reads as follows.

THEOREM 2.1.2. *Let* k *be a number field with ring of integers* O_k. *Let* S *be a finite set of places of* k. *For each* $v \in S$ *let* $L_{v,i}$ $(0 \leq i \leq n)$ *be* $n + 1$ *linearly independent linear forms in* $n + 1$ *variables, with algebraic coefficients. If* $s = (s_0, \ldots, s_n) \in O_k^{n+1}$, *let*

$$\text{size}(s) = \max_{\substack{v \mid \infty \\ 0 \leq i \leq n}} \|s_i\|_v.$$

Fix $\epsilon > 0$. *Let* Q *be the set of all* $s \in O_k^{n+1}$ *satisfying,*

$$\prod_{v \in S} \prod_{i=0}^{n} \|L_{v,i}(s)\|_v < \text{size}(s)^{-\epsilon}.$$

Then Q *is contained in a finite union of hyperplanes of* k^{n+1}.

§2. A Reformulation of Schmidt's Theorem

We start by rephrasing Theorem 2.1.2 as follows:

THEOREM 2.2.1. *Let* k, O_k, S, $L_{v,i}$, *and* ϵ *be as in the previous section. Let* Q' *be the set of all points* $P \in \mathbf{P}^n(k)$ *which can be written as* $P = (s_0, \ldots, s_n)$ *with* $s_i \in O_k$ *and which satisfy the inequality,*

$$\prod_{v \in M_k} \prod_{i=0}^{n} \|L_{v,i}(s)\|_v < H(P)^{-\epsilon}.$$

Then Q' *lies in a finite union of hyperplanes of* $\mathbf{P}^n$.

Theorem 2.2.1 implies Theorem 2.1.2 easily; indeed, if $P = (s_0, \ldots, s_n)$ with $s_i \in O_k$, then trivially

$$\text{size}(s)^{[k:\mathbf{Q}]} \gg H(P).$$

The converse is a consequence of the following lemma:

LEMMA 2.2.2. *Any point* $P \in \mathbf{P}^n(k)$ *has homogeneous coordinates* $(s_0, \ldots, s_n)$ *such that* $s_i \in O_k$ *and*

$$\text{size}(s) \ll H(P).$$

(The constant in $\ll$ does not depend on P.)

PROOF: Choose coordinates $(s_0, \ldots, s_n)$ for P such that:

  (1). The coordinates are integers (not merely S-integers).
  (2). They are quasi-relatively prime (i. e. the ideal $(s_0, \ldots, s_n)$ has norm bounded by a constant c_1 depending only on k).
  (3). For all archimedean places v,

$$\max_{0 \leq i \leq n} \|s_i\|_v \geq c_2,$$

  where c_2 depends only on k.

Condition (1) is trivial to satisfy; condition (2) is feasible since the class number is finite. Choose $(s_0, \ldots, s_n)$ satisfying (1) and (2). Scaling these coordinates by a unit will not affect these first two conditions. By the proof of the Dirichlet unit theorem, the image of the units is a full lattice in the usual subspace of the logarithmic space; therefore a unit u exists such that $(us_0, \ldots, us_n)$ satisfies (3).

With this choice of coordinates, the proof is complete since,

$$\text{size}(s) \leq \frac{c_1 H(P)}{c_2^{[k:\mathbf{Q}]}}.$$ □

A straightforward calculation will show that Theorem 2.2.1 implies Roth's theorem.

REMARK 2.2.3. In Theorem 2.2.1, we may assume that the coefficients of the linear forms all lie in k. Indeed, let $\text{RS}_{E/k}$ denote the assertion that the Roth-Schmidt theorem holds for all collections of hyperplanes defined over E, with P varying over $\mathbf{P}^n(k)$. Then Theorem 2.3.3 is the assertion that $\text{RS}_{k/k}$ holds for all number fields k, whereas Schmidt's theorem is $\text{RS}_{\overline{\mathbf{Q}}/k}$. Trivially, however, we have

$$\text{RS}_{\overline{\mathbf{Q}}/k} = \text{``}\text{RS}_{E/k} \text{ for all } E \supseteq k.\text{''}$$

since we can take E to be a number field containing all coefficients of the $H_{v,i}$. Our assertion now follows from the fact that $\text{RS}_{E/E}$ implies $\text{RS}_{E/k}$ if E is normal over k; we leave this last step to the reader.

Finally, we restate the theorem in terms of Weil functions, using (1.3.5). We say that a set of hyperplanes of $\mathbf{P}^n$ is in <u>general</u> <u>position</u> if all sets of up to $n+1$ of the hyperplanes are linearly independent.

THEOREM 2.2.4. *For each* $v \in S$, *let* $H_{v,1}, \ldots, H_{v,m}$ *be* m *hyperplanes of* $\mathbf{P}^n$ *in general position, defined over a number field* k. *Let* $\lambda_{v,i}$ *be the corresponding Weil functions, and fix* $\epsilon > 0$. *Then all* $P \in \mathbf{P}^n$ *for which*

$$\sum_{v \in S} \sum_{i=1}^{m} \lambda_{v,i}(P) > (n+1+\epsilon)h(P),$$

lie in finitely many hyperplanes.

PROOF: For $m \leq n+1$, this follows trivially from Lemma 1.3.3 (a) and (g). For $m > n+1$, we note that $\lambda_{v,i}$ is large only if P is close to H_i in the v-adic topology. But since the $H_{v,i}$ are in general position, at most n of them may pass through a given point, so P can be v-adically close to at most n of the $H_{v,i}$ for any given v. For the other hyperplanes, $\lambda_{v,i}$ is bounded, so the theorem follows from applying Theorem 2.2.1 to all possible subcollections of the $\lambda_{v,i}$ with at most $n+1$ hyperplanes at each v. □

§3. Linear Equations in Units

The main result of this section is due to van der Poorten [vdP], who generalized an idea of Schlickewei [Schl 2] to obtain a general theorem.

THEOREM 2.3.1. *Let* k *be a number field and let* $n > 1$ *be an integer. Let* Γ *be a finitely generated subgroup of* k^*. *Then all but finitely many solutions of the equation,*

$$(2.3.2) \qquad u_0 + u_1 + \cdots + u_n = 1, \qquad u_i \in \Gamma,$$

lie in one of the diagonal hyperplanes H_I *defined by the equation* $\sum_{i \in I} x_i = 0$, *where* I *is a subset of* $\{0, \ldots, n\}$ *with at least two, but no more than* n, *elements.*

REMARK. This theorem can be generalized to allow k to be any finitely generated field of characteristic zero, merely by specializing k to a number field in such a way that $\Gamma \cdot \{\pm 1\}$ injects. M. Laurent [Lau] has generalized this to allow Γ to be a subgroup of $\mathbf{C}^*$ finitely generated over its torsion subgroup.

PROOF: We may assume that Γ is the group of S-units of k. Let a solution $[u_0 : \cdots : u_n]$ of (2.3.2) define a point $P \in \mathbf{P}^n$. Also, we define $n + 2$ hyperplanes H_i by $X_0 = 0, \ldots, X_n = 0$, $X_0 + \cdots + X_n = 0$. Then by (1.3.5),

$$\lambda_{v,i}(P) = -\log \frac{\|u_i\|_v}{\max(\|u_0\|_v, \ldots, \|u_n\|_v)}, \qquad 0 \le i \le n;$$

$$\lambda_{v,n+1}(P) = \log \max(\|u_0\|_v, \ldots, \|u_n\|_v).$$

Since the u_i are units, we have $\lambda_{v,i}(P) = 0$ for all $v \notin S$; by (1.3.4),

$$\sum_{v \in S} \sum_{i=0}^{n+1} \lambda_{v,i}(P) = (n+2)h(P).$$

By Theorem 2.2.4, all solutions of (2.3.2) must therefore lie in a finite set of hyperplanes.

By induction on n, it can be seen that the only subspaces containing infinitely many solutions are the diagonal subspaces. Following [Gr 1], we first show that any infinite family of solutions $[u_0, \ldots, u_n]$ of the unit equation,

$$(2.3.3) \qquad \sum_{i=0}^{n} a_i u_i = 1$$

has an infinite subsequence of solutions in which some u_i is constant. We already know that all such solutions lie in a finite collection of hyperplanes; after subsequencing, we may assume that they all lie in one hyperplane given by the equation,

$$\sum_{i=0}^{n} b_i u_i = 0,$$

where $b_i \in k$ are constants, not all zero. Assuming $b_n \ne 0$, we can eliminate u_n from (2.3.3), giving

$$(2.3.4) \qquad a_0' u_0 + \cdots + a_{n-1}' u_{n-1} = 1.$$

Rearranging indices, we may assume that $a_i \ne 0$ for $0 \le i \le m$ and $a_i' = 0$ for $i > m$, for some $m < n$. If our family of solutions gives an infinite family of solutions $(u_0, \ldots, u_m)$ of (2.3.4), then we proceed by induction on n; otherwise, there is an infinite subsequence with $u_0, \ldots, u_m$ constant.

Now, returning to (2.3.2), we pass to an infinite subsequence and rearrange indices, so that $u_0, \ldots, u_n$ are constant and m is maximal. Then, by maximality and by what we have just proved, we have $u_0 + \cdots + u_m = 1$. Thus the sequence is contained in a diagonal hyperplane. □

Actually, one can conclude much more—in fact, such infinite families are restricted to finite unions of linear subspaces of dimension $\leq [n/2]$. For example, if $n = 4$, then we have the diagonal subspace $u_0 + u_1 + u_2 = 0$, containing infinitely many solutions (in a sufficiently large field). In this subspace all solutions are limited to finitely many lines ($\subseteq \mathbf{P}^4$) since all solutions must have $u_0 + u_1 + u_2 = 0$, $u_3 + u_4 = 1$. But such lines are not geometric: the set of lines increases without bound as k and S increase. Therefore, we will stick with the statement using diagonal hyperplanes.

§4. A Theorem on Integral Points

Theorem 2.3.1 of the last section implies a theorem on integral points which partially extends Siegel's theorem to higher dimensional varieties. We say "partially extends" because, when reduced to the case of curves, the theorem gives only a weak version of Siegel's theorem. Before stating the theorem, however, it is necessary to recall some definitions from algebraic geometry.

Let V be a nonsingular variety defined over a field of characteristic zero. Its group of divisors $\operatorname{Div} V$ has two quotient groups which are of interest, the Picard group and the Néron-Severi group. Respectively, these groups are the group of divisors modulo linear equivalence and algebraic equivalence. It is known that linearly equivalent divisors are also algebraically equivalent, so there exists a surjection,

$$\operatorname{Pic} V \twoheadrightarrow \operatorname{NS} V.$$

The kernel of this map is called the Picard variety and is denoted Pic^0. It has the structure of an abelian variety. The image is a finitely generated abelian group. Its free rank is called the Picard number and denoted ρ.

If k is a number field, the group of divisors defined over k is a subgroup of the group of divisors defined over $\mathbf{C}$, so $\operatorname{NS}(V)$ and $\operatorname{Pic}^0(V)$ have subgroups of divisor classes defined over k. In the case of $\operatorname{Pic}^0(V)$, the above subgroup coincides with the group of k-rational points of the abelian variety $\operatorname{Pic}^0(V)$. Therefore it is a finitely generated subgroup of $\operatorname{Pic}^0(V)$, by the Mordell-Weil theorem.

It is now possible to state the major theorem of this chapter.

THEOREM 2.4.1. *Let* r *be the rank of the group of* k-*rational points of the variety* $\mathrm{Pic}^0(V)$. *Assume* D *is a divisor which is a sum of at least* $\dim V + \rho + r + 1$ *distinct prime divisors* D_i, *all defined over* k. *Then all sets of* D-*integral points of* V *are degenerate.*

REMARK. Recall (Definition 1.4.13) that a set is degenerate if it lies in a proper Zariski-closed subset.

PROOF: By the above comments, the group of divisor classes having a representative defined over k has rank at most $r + \rho$. Therefore there must exist at least $\dim V + 1$ linearly independent relations between the divisors D_i; i. e. $\dim V + 1$ divisors which can be written as sums of the D_i, are linearly equivalent to zero, and are independent as elements of $\mathrm{Div}\, V$. These divisors can therefore be written as (f_i), where f_i are multiplicatively independent rational functions on V. Since the poles of f_i are in D, given a set of integral points there exists $a \in k$ such that af_i takes on integral values at those integral points (Lemma 1.4.6). But the same reasoning holds for $1/f_i$, so values of each f_i may lie in only finitely many cosets of the units at each integral point.

Since there are more functions f_i than $\dim V$, there exists a polynomial $p(f_1, \ldots, f_{d+1})$ which vanishes identically on V. Choose such a polynomial with a minimal number of terms. Monomials of this polynomial also must lie in only finitely many cosets of the units. Thus, at integral points, p becomes a relation of the form (2.3.2). Solutions of (2.3.2) fall into two categories. First there are the diagonal solutions, which satisfy $\sum u_i = 0$ for some subset of the monomials u_i indexed by $i \in I$. These relations determine subvarieties of V by the minimality assumption on p. These subvarieties are finite in number (bounded by two raised to the power of the number of monomials in p). This leaves nontrivial solutions of (2.3.2)—a finite number of points. Normally each point would determine a finite number of points on V, unless p does not involve all of the functions f_i, in which case each point could give a subvariety.

In either case, the proof of the theorem is complete. □

REMARK 2.4.2. If each nontrivial solution of (2.3.2) can only come from finitely many points on V, then the higher dimensional part of the Zariski-closed subsets of V would be independent of k. This holds for example if $V = \mathbf{P}^n$ and the H_i are hyperplanes, and it is plausible that it holds much more generally (cf. (3.4.3)).

The special case when Pic^0 vanishes is of particular interest. This is the case, for example, when $\dim V > 1$ and V is a complete intersection. Then the hypotheses of the theorem are satisfied when D has a certain number of components, independent of the field of definition. In this case also the following corollaries hold.

COROLLARY 2.4.3. $\mathbf{P}^n$ has no non-degenerate sets of D-integral points, provided D has at least $n + 2$ distinct components.

COROLLARY 2.4.4. For almost all minimal K3 surfaces V, if D is a divisor on V having at least four components, then the same conclusion holds.

Indeed, $\rho = 1$ for $\mathbf{P}^n$ and for almost all K3 surfaces (see [Sha, Ch. IX] for the latter).

As a further application, consider the case of abelian varieties. In this case Pic^0 does not vanish—in fact, it is isomorphic to the dual of the variety. Furthermore, $\mathrm{Pic}^0(V)$ is also isomorphic to the quotient of V by a finite group, so if k is large enough to contain this morphism, then the rank of the group of k-rational points of Pic^0 is not smaller than the same rank for V. Since $V \cong \mathrm{Pic}^0(\mathrm{Pic}^0(V))$, these ranks are the same. Thus, we have the following corollary.

COROLLARY 2.4.5. Assume V is an abelian variety, k is sufficiently large (as above), r is the rank of the group of k-rational points on V, and D has at least $\rho + r + \dim V + 1$ distinct prime divisors defined over k. Then all sets of D-integral points of V are degenerate.

This provides a partial answer to a conjecture of Lang, which states that all sets of D-integral points of V are <u>finite</u>, if V is an abelian variety and D is any ample divisor [L 1].

As a final application, we note that one can eliminate the dependence on ρ and on the rank of $\mathrm{Pic}^0(V)(k)$ by requiring that the prime divisors D_i be hyperplane sections.

THEOREM 2.4.6.

(a). Let V be a variety defined over a number field k. Let D be a divisor which is a sum of $\mathrm{Div}\, V + 2$ nonredundant hypersurface sections of V (in some fixed projective embedding). Then all sets of D-integral points of V are degenerate.

(b). *(M. Green, [Gr 1]). Let V and D be as above. Then all holomorphic maps $\mathbf{C} \to V$ whose image omits D must lie in some hypersurface section.*

PROOF: (a) follows from a simplification of the argument used to prove Theorem 2.4.1. (b) follows from the same argument as (a), replacing Theorem 2.3.1 with the Borel lemma:

THEOREM (BOREL, [Bo]). *Let $g_1, \ldots, g_n$ be entire functions satisfying,*

$$e^{g_1} + \cdots + e^{g_n} = 1.$$

Then some g_i is constant.

$\square$

REMARK 2.4.7. Thus we see another example (cf. Remark 1.4.12) of how a holomorphic map behaves analogously to an infinite set of integral points. It was the remarkable similarity between parts (a) and (b) of Theorem 2.4.6 that led to the main conjecture of this work, which will be the subject of the next chapter.

§5. Arithmetic Distance Functions and Schmidt's Theorem

With an appropriate notion of Weil functions attached to subvarieties of arbitrary codimension, it is possible to restate Theorem 2.2.4 without requiring the hyperplanes to lie in general position.

DEFINITION 2.5.1 ([Sil 4]). *Let C be an effective cycle on a nonsingular variety V. Assume C can be written as an intersection $C = \bigcap D_i$ of a finite number of effective divisors. Moreover, we assume that this relation holds with multiplicities as well; i. e. if I_C and I_{D_i} are the associated sheaves of ideals, then $I_C = \sum I_{D_i}$. Let k be a number field such that V, C, and D_i are defined over k and let v be a valuation of k. Then an* <u>arithmetic distance function</u> *relative to C is a function,*

$$\lambda_{C,v} : V(\overline{k}_v) \setminus |C| \to \mathbf{R}$$

defined by letting

$$\lambda_{C,v}(P) = \min_i \lambda_{D_i,v}(P)$$

for some choice of Weil functions λ_{D_i}. We let $\lambda_{D_i,v}(P) = \infty$ if $P \in D_i$.

As with Weil functions, we can define a <u>global</u> arithmetic distance function using global Weil functions in the above formula. By choosing an embedding of $\overline{k}$ in $\overline{k}_v$, we can consider $\lambda_{C,v}$ as a function on $V(\overline{k}) \setminus |C|$. Often we will omit the subscript v. We are only interested in arithmetic distance functions up to a constant (or, in the global case, up to a set of constants c_v which equal zero for almost all v). By additivity, these definitions can be extended to the group of all cycles on V.

Silverman also shows that arithmetic distance functions satisfy all but part (g) of Lemma 1.3.3. In particular, let $\phi : V \to W$ be a morphism of nonsingular varieties. Let C be a cycle on W, and let $\phi^* C$ denote the corresponding cycle on V. Then $P \mapsto \lambda_C(\phi(P))$ is an arithmetic distance function relative to $\phi^* C$ on V. This implies,

LEMMA 2.5.2. *If C is a cycle on a nonsingular variety W, and $\pi : V \to W$ is a blowing up of C with exceptional <u>divisor</u> $E = \pi^* C$, then the arithmetic distance function on W,*

$$\lambda_C(P) = \lambda_E(\pi^{-1}(P))$$

corresponds to the Weil function on V.

DEFINITION 2.5.3. *Let $H_1, \ldots, H_m$ be a set of hyperplanes in $\mathbf{P}^n$ (which need not lie in general position). Then the <u>associated cycle</u> of $\{H_i\}$ is the cycle $\sum n_i C_i$ such that:*

(a). *The set of components C_i is the set of nonempty linear subspaces of $\mathbf{P}^n$ which can be written as an intersection of one or more of the hyperplanes H_i.*

(b). *The multiplicities n_i satisfy the equation,*

$$(2.5.4) \qquad\qquad \sum n_j = \operatorname{codim} C_i,$$

where the sum is over all j such that $C_j \supseteq C_i$. In particular, $n_i = 1$ if C_i is a hyperplane and $n_i \leq 0$ otherwise. If $\{H_i\}$ are in general position, then this cycle equals $\sum H_i$.

LEMMA 2.5.5. *Let $H_1, \ldots, H_n$ and $C = \sum n_i C_i$ be as above. Let λ_{H_i} be arithmetic distance functions (=Weil functions) for the H_i and let λ_C be an arithmetic distance function for C. Then*

$$\lambda_C(P) = \max_T \sum_{H \in T} \lambda_H(P) + O(1)$$

where the maximum is over all linearly independent subsets T of $\{H_1, \ldots, H_m\}$.

PROOF: Blow up $\mathbf{P}^n$ along each component C_i of C of codimension ≥ 2, obtaining a map $\pi: V \to \mathbf{P}^n$ such that the inverse image $\pi^*(C_i)$ is a divisor for all i. We may take this blowing-up to be minimal, so that each component of the exceptional divisor of π lies over some C_i. Let $E_i^{\#}$ denote the strict transform of C_i; i. e. the prime divisor on V lying over C_i. Then the support of $\pi^*(C)$ is contained in $\bigcup E_i^{\#}$; in fact,

$$(2.5.6) \qquad \pi^* C_i = \sum_{j \mid C_j \subseteq C_i} E_j^{\#},$$

so that by (2.5.4),

$$(2.5.7) \qquad \pi^*(C) = \sum (\operatorname{codim} C_i) \cdot E_i^{\#}.$$

Thus the problem is reduced to a statement of Weil functions, so it suffices to prove the following identity of divisors:

$$\pi^*(C) = \max_T \sum_{H \in T} \pi^*(H).$$

But by (2.5.6), the coefficient of $E_i^{\#}$ in $\sum_{H \in T} \pi^*(H)$ is just the number of hyperplanes in T containing C_i. This is always $\leq \operatorname{codim} C_i$, with equality for some suitable T. Thus, by (2.5.7), the lemma is proved. $\square$

As a corollary, we obtain the following version of Schmidt's theorem (2.2.4):

THEOREM 2.5.8. *Let $H_1, \ldots, H_m$ be hyperplanes in $\mathbf{P}^n$ with associated cycle $C = \sum n_i C_i$, and let λ_C be an arithmetic distance function of C. Let k be a number field over which $H_1, \ldots, H_m$ are defined, and let S be a finite set of places of k. Let $\epsilon > 0$. Then all points $P \in \mathbf{P}^n(k)$ for which,*

$$(2.5.9) \qquad \sum_{v \in S} \lambda_{C,v}(P) > (n + 1 + \epsilon)h(P)$$

lie in a finite union of hyperplanes of $\mathbf{P}^n$.

§6. Exceptional Hyperplanes

Let k, S, ϵ, H_1, ..., H_m, and C be as in the previous section. For the remainder of this chapter, the word "dense" will refer to the topology whose closed subsets are the finite unions of linear subspaces.

DEFINITION 2.6.1. *An <u>exceptional subspace</u> relative to k, S, ϵ, etc. is a maximal subspace H containing a dense subset of points P for which (2.5.9) holds.*

In §3, we have already characterized the exceptional subspaces in the special case where $\epsilon = 1$, $m = n + 2$, and H_1, ..., H_m lie in general position. By a linear transformation, we may assume these planes to be $H_i = \{x_i = 0\}$ for $0 \le i \le n$ and $H_{n+1} = \{x_0 + \cdots + x_n = 0\}$. Then the exceptional subspaces are the <u>diagonal</u> <u>hyperplanes</u> H_I defined by,

$$\sum_{i \in I} x_i = 0,$$

where I is a subset of $\{0, 1, \ldots, n\}$ with at least two, but not more than n, elements. Geometrically, such hyperplanes are spanned by two subspaces $\bigcap_{i \in I} H_i$ and $\bigcap_{i \notin I} H_i$. Thus, in this case, the set of exceptional subspaces is finite and independent of k and S.

In more general cases, however, there are exceptional subspaces not of the above form, although I still expect that there be at most a finite number of exceptional subspaces, independently of k, S, and ϵ.

EXAMPLE 2.6.2. Let H_1, ..., H_9 be planes in $\mathbf{P}^3$ in general position. Let P_0 be the point of intersection of H_1, H_2, and H_3. Likewise, let $P_1 = H_4 \cap H_5 \cap H_6$ and $P_2 = H_7 \cap H_8 \cap H_9$. Let Y be the plane defined by P_0, P_1, and P_2. Assign homogeneous coordinates on Y such that $P_0 = [1 : 0 : 0]$, $P_1 = [0 : 1 : 0]$, and $P_2 = [0 : 0 : 1]$.

Assume H_1, ..., H_9 are defined over a number field k, and let S be a finite set of places of k containing the archimedean places, and having at least three places v_0, v_1, and v_2. In the following we will use the notation $v(x) = -(1/[k : \mathbf{Q}]) \log \|x\|_v$. By the Dirichlet unit theorem, S-units u of k span a lattice in the subspace $\sum_{v \in S} v(u) = 0$ of $\mathbf{R}^{\#S}$. Thus there exists a constant c independent of $x > 0$, such that the conditions,

$$-x - c \le v_i(u_i) \le -x + c \qquad i = 1, 2;$$
$$x - c \le v_0(u_i) \le x + c \qquad i = 1, 2;$$
$$(2.6.3) \qquad |v_1(u_2)| \le c;$$
$$|v_2(u_1)| \le c;$$
$$|v(u_i)| \le c \qquad \text{for all other } v \in S, \ i = 1, 2.$$

are satisfied for at least one pair of S-units u_1, u_2 of k. As x varies we will obtain infinitely many pairs (u_1, u_2).

If L_0 denotes the line $x_0 = 0$ at infinity, then a Weil function is

$$\lambda_{L_0}([x_0 : x_1 : x_2]) = \max(v(\frac{x_0}{x_0}), v(\frac{x_0}{x_1}), v(\frac{x_0}{x_2}))$$
$$= v(x_0) - \min(v(x_0), v(x_1), v(x_2)).$$

Hence, the arithmetic distance function $\lambda_{P_0, v}$ relative to P_0 is

$$\lambda_{P_0} = \min(\lambda_{L_1}, \lambda_{L_2})$$
$$= \min(v(x_1), v(x_2)) - \min(v(x_0), v(x_1), v(x_2)).$$

Thus, the points $U = [1 : u_1 : u_2]$ of $(2.6.3)$ have,

$$\lambda_{P_i, v}(U) = \begin{cases} x + O(1) & \text{if } v = v_i; \\ O(1) & \text{otherwise.} \end{cases}$$

In any case,

$$\sum_{v \in S} \lambda_{P_i, v}(U) = x + O(1) \qquad \text{for } i = 0, 1, 2;$$
$$\sum_{v \in S} \lambda_{H_i, v}(U) = x + O(1) \qquad \text{for } i = 1, 2, \ldots, 9.$$

Then, since $h(U) = 2x + O(1)$, we have,

$$\sum_{v \in S} \sum_{i=1}^{9} \lambda_{H_i, v}(U) = \tfrac{9}{2} h(U) + O(1).$$

As x varies, the U_x converge to P_i in the v_i topology, so they must be dense. Thus Y is an exceptional subspace of $\mathbf{P}^3$ relative to $H_1, \ldots, H_9$.

DEFINITION 2.6.4. *Let $H_1, \ldots, H_m$ be hyperplanes in $\mathbf{P}^n$ and assume that at least $n + 1$ of them are in general position relative to each other. If T is a subset of n linearly independent elements of $\{H_i\}$, then let P_T be their point of intersection. Then the generalized diagonals of $\mathbf{P}^n$ relative to $H_1, \ldots, H_m$ are the proper linear subspaces of $\mathbf{P}^n$ spanned by some subset of the P_T.*

From the geometric description of the diagonals one can see that diagonal subspaces are generalized diagonals. For example, $H_1 \cap \cdots \cap H_i$ is spanned by the points $H_1 \cap \cdots \cap \widehat{H_{i+1}} \cap \cdots \cap H_{n+1}, \ldots, H_1 \cap \cdots \cap H_n$, and likewise for the other spanning subspace. On the other hand, the subspace in the above example is not of diagonal type.

As in the example, it is possible to show that the generalized diagonals are exceptional subspaces. It is also true, in dimensions ≤ 4, that all exceptional subspaces are generalized diagonals. In general, though, this is false: there is an example of an exceptional subspace $\mathbf{P}^4 \subseteq \mathbf{P}^{13}$ which is not a generalized diagonal. These results will be the subject of a subsequent paper.

Chapter 3
A Correspondence with Nevanlinna Theory

In this chapter we will examine, explicitly, what has been discussed in the Introduction, namely that the diophantine behavior of curves closely parallels the behavior of holomorphic maps to the corresponding Riemann surfaces. We then make a diophantine conjecture for rational points on higher dimensional varieties, based on the hope that the above correspondence for curves is valid also for higher dimensional varieties. This hope is already supported by Remarks 1.4.12 and 2.4.7.

For curves, what has been noticed is that an algebraic curve has infinitely many points over a sufficiently large ring of S-integers of a number field, exactly when the corresponding Riemann surface admits a non constant transcendental holomorphic map. On the arithmetical side, all such theorems (except the Mordell conjecture) can be proved using Roth's theorem; on the analytic side, all these statements follow from the branch of complex analysis known as Nevanlinna theory.

Like Roth's theorem, Nevanlinna theory is a theory involving approximations. C. Osgood ([**Os 1**]) has observed that there is an even deeper correspondence; in particular, there is a "2" in Nevanlinna theory that plays the same role as the "$2 + \epsilon$" appearing in Roth's theorem. He has used this to translate Roth's proof into a "sometimes effective" proof of Nevanlinna's Second Main Theorem. In this chapter we will consider this correspondence and how it might benefit number theory.

In Section 1 we introduce Nevanlinna theory in the case of meromorphic functions on $\mathbf{C}$. In Section 2 we describe the above correspondence in greater detail, giving a dictionary for translating between the two (Table 3.1). With this dictionary, the statements of the theorems in Nevanlinna theory and diophantine approximations become virtually identical.

In Section 3 we discuss the defect $\delta(a)$, as translated from Nevanlinna theory, and show how it can be used to deduce Seigel's theorem (on integral points on rational varieties) from Roth's theorem by a purely formal argument. We note that E. Reyssat [**Rey**] has already observed the similarity between arithmetic statements on curves (such as finiteness for the Thue and unit equations) and the corresponding analytic theorems.

Since these two areas of mathematics have developed totally independently of one another, it is highly improbable that their current states of development be exactly the same. In fact, Nevanlinna theory is currently ahead of number theory. Therefore, it makes sense to attempt to translate recent ideas in Nevanlinna theory into the arithmetical case. This is the subject of Section 4.

Finally, Section 5 concludes the chapter with some examples, mostly derived from Nevanlinnna theory, which show the necessity of the smoothness and nondegeneracy conditions of the conjecture of Section 4.

§1. Introduction to Nevanlinna Theory

Nevanlinna theory is the study of distribution of values of meromorphic functions. Hadamard made the first discovery in this direction. He observed that, given an entire function, the maximum modulus $|f(x)|$ inside the circle $|z| = r$, as a function of r, grows at least as fast as the exponential of the number of zeroes of f in that circle. For a function with poles, the maximum modulus is no longer defined, though. In the 1920's, however, R. Nevanlinna found that the maximum modulus could be replaced by the characteristic function as described below; he was then able to prove two beautiful theorems relating the growth of a meromorphic function to the number of zeroes inside the circle of radius r.

To describe those theorems, let $f\colon \mathbf{C} \to \mathbf{C}$ be a meromorphic function. Make the following definitions:

$$\log^+ x = \max(\log x, 0)$$

$$m(a, r) = \int_0^{2\pi} \log^+ \left| \frac{1}{f(re^{i\theta}) - a} \right| \frac{d\theta}{2\pi} \qquad \text{(proximity function)}$$

$$m(\infty, r) = \int_0^{2\pi} \log^+ |f(re^{i\theta})| \frac{d\theta}{2\pi}$$

$n(a, r) = $ no. of zeroes of $f - a$ inside the circle of radius r

$n(\infty, r) = $ no. of poles of f inside the circle of radius r

$$N(a, r) = \int_0^r n(a, s) \frac{ds}{s} \qquad \text{(counting function)}$$

$$= \sum_{|w|<r} \operatorname{ord}_w^+ (f - a) \log \frac{r}{|w|}.$$

$$N(\infty, r) = \int_0^r n(\infty, s) \frac{ds}{s} = \sum_{|w|<r} \operatorname{ord}_w^+ \frac{1}{f} \log \frac{r}{|w|}.$$

$$T(r) = \int_0^{2\pi} \log^+ |f(re^{i\theta})| \frac{d\theta}{2\pi} + N(\infty, r)$$

(characteristic function)

The Nevanlinna inequality states,

$$N(a, r) < T(r) + O(1),$$

where the constant in $O(1)$ may depend on a and f but not on r. This is a generalization of Hadamard's observation. The above inequality is not an equality; for example, the function e^z has no zeroes but grows rapidly. The correction term $m(a, r)$ makes this into an equality. This gives Nevanlinna's First Main Theorem,

$$N(a, r) + m(a, r) = T(r) + O(1).$$

In the example of e^z, $m(0, r)$ is large because e^z has a large number of very small values.

The Second Main Theorem is the assertion that for any distinct complex numbers $a_1, \ldots, a_n$,

$$\sum_{i=1}^{n} m(a_i, r) \leq 2T(r) - N_1(r) + O(\log rT(r)). \qquad //$$

The constant in $O()$ depends on the function f and on $a_1, \ldots, a_n$. Following Weyl, we let $//$ indicate that the inequality holds outside a set of bounded measure. $N_1(r)$ counts the zeroes of $f'(z)$ in the same way as $N(a, r)$ counts the zeroes of $f(z) - a$; in particular it is nonnegative and we will ignore it until Chapter 5.

In our example e^z, we have $m(0, r) = m(\infty, r) = T(r)$, so that for all finite $a \neq 0$, we must then have $m(a, r)$ small. With the First Main Theorem, this gives a lower bound on most of the $N(a, r)$, giving a converse to the upper bound provided by the First Main Theorem alone.

Now define the Nevanlinna defect as,

$$\delta(a) = \liminf_{r \to \infty} \frac{m(a, r)}{T(r)}$$

The Nevanlinna defect relation states that if f is a non constant meromorphic map, then

$$\sum_{a \in \mathbf{C}} \delta(a) \leq 2.$$

§2. The Correspondence in the Case of One Variable

Throughout the remainder of this chapter, let k be a number field and S a finite set of places of k containing the set of archimedean places. We assume all varieties, divisors, etc. to be defined over k.

As discussed in the introduction, we want to let a meromorphic function correspond to an infinite set $\mathcal{R}$ of elements of k. To do this, we split f, by contraction, into an infinite number of maps from the closed unit disc $\mathbf{D}$ into $\mathbf{C}$. One such map corresponds to one element $b \in \mathcal{R}$. Absolute values of f on the boundary of the disk correspond to absolute values of b at places in S; zeroes and poles of f in the interior of $\mathbf{D}$ correspond to the order of b relative to places outside of S. Table 3.1 summarizes the above identifications, as well as the first and second main theorems.

Now let us translate the counting function. Analytically, it is defined as,

$$N(a, r) = \sum_{|w| < r} \operatorname{ord}_w^+ (f - a) \log \frac{r}{|w|}.$$

Performing the above substitutions, changing $r/|w|$ to $N\wp$, we have

$$N(a, b) = \frac{1}{[k : \mathbb{Q}]} \sum_{\wp \notin S} \operatorname{ord}_\wp^+ (b - a) \log N\wp$$

$$= \frac{1}{[k : \mathbb{Q}]} \sum_{\wp \notin S} \log^+ \left\| \frac{1}{b - a} \right\|_\wp.$$

Similarly, the translated proximity function is,

$$m(a, b) = \frac{1}{[k : \mathbb{Q}]} \sum_{v \in S} \log^+ \left\| \frac{1}{b - a} \right\|_v.$$

Both the counting function and the proximity function depend on S; sometimes we will write $N_S(a, b)$ and $m_S(a, b)$ to emphasize this fact.

The characteristic function translates into,

$$\frac{1}{[k : \mathbb{Q}]} \sum_{v \in S} \log^+ \|b\|_v + \frac{1}{[k : \mathbb{Q}]} \sum_{v \notin S} \log^+ \|b\|_v = h(b),$$

the logarithmic height. Both measure the complexity of the number (or function).

Nevanlinna Theory	Roth's Theorem

$$f: \mathbf{C} \to \mathbf{C}, \quad \text{non constant} \qquad \{b\} \subseteq k, \quad \text{infinite}$$

r	b
θ	$v \in S$
$|f(re^{i\theta})|$	$\|b\|_v, \quad v \in S$
$\text{ord}_z f$	$\text{ord}_v f, \quad v \notin S$
$\log \dfrac{r}{|z|}$	$\log N_v$

Characteristic function   **Logarithmic height**

$$T(r) = \int_0^{2\pi} \log^+ |f(re^{i\theta})| \frac{d\theta}{2\pi} + N(\infty, r) \qquad h(b) = \frac{1}{[k:\mathbf{Q}]} \sum_v \log^+ \|b\|_v$$

Proximity function

$$m(a,r) = \int_0^{2\pi} \log^+ \left| \frac{1}{f(re^{i\theta}) - a} \right| \frac{d\theta}{2\pi} \qquad m(a,b) = \frac{1}{[k:\mathbf{Q}]} \sum_{v \in S} \log^+ \left\| \frac{1}{b-a} \right\|_v$$

Counting function

$$N(a,r) = \sum_{|w|<r} \log \frac{r}{|w|} \qquad N(a,b) = \frac{1}{[k:\mathbf{Q}]} \sum_{v \notin S} \log^+ \left\| \frac{1}{b-a} \right\|_v$$

First Main Theorem   **Property of heights**

$$N(a,r) + m(a,r) = T(r) + O(1) \qquad N(a,b) + m(a,b) = h(b) + O(1)$$

Second Main Theorem   **Conjectured refinement of Roth**

$$\sum_{i=1}^m m(a_i, r) \le 2\,T(r) - N_1(r)$$
$$+ O(r \log T(r)) \quad // \qquad \sum_{i=1}^m m(a_i, b) \le 2\,h(b) + O(\log h(b))$$

Defect $\delta(a) = \liminf\limits_{r \to \infty} \dfrac{m(a,r)}{T(r)}$ $\delta(a) = \liminf\limits_{b} \dfrac{m(a,b)}{h(b)}$

Defect Relation $\sum\limits_{a \in \mathbf{C}} \delta(a) \le 2$ $\sum\limits_{a \in k} \delta(a) \le 2$ **Roth's theorem**

Jensen's formula   **Artin-Whaples Product Formula**

$$\log |c_\lambda| = \int_0^{2\pi} \log |f(re^{i\theta})| \frac{d\theta}{2\pi} \qquad \sum_v \log \|b\|_v = 0$$
$$+ N(\infty, r) - N(0, r)$$

Table 3.1. The Dictionary in the One Dimensional Case.

A first step in the proof of the First Main Theorem is Jensen's formula,

$$\log |c_\lambda| = \int_0^{2\pi} \log |f(re^{i\theta})| \frac{d\theta}{2\pi} + N(\infty, r) - N(0, r).$$

(c_λ is the leading coefficient in the Laurent expansion.) The left-hand side has no obvious counterpart, but the right-hand side translates into,

$$\sum_{v \in S} \log \|b\|_v + \sum_{v \notin S} \log^+ \|b\|_v - \sum_{v \notin S} \log^+ \left\| \frac{1}{b} \right\|_v = \sum_v \log \|b\|_v.$$

Translating the left-hand side into 0 gives the Artin-Whaples product formula.

Note that the First Main Theorem in the number field case is actually property (1.2.9b) of heights; i. e. that linearly equivalent divisors give rise to equivalent height functions.

To translate the Second Main Theorem, we take a weaker version, namely,

$$(3.2.1) \qquad \sum_{i=1}^n m(a_i, r) < (2 + \epsilon) T(r) \qquad\qquad //_\epsilon$$

This is already the same as Roth's Theorem (2.2.4), provided that all $a_i \in k$. If not, we use the following argument, due to Lang. Add elements to the set $\{a_i\}$ so that the divisor $D = \sum [a_i]$ is defined over k. Then we can define Weil functions for D over k; (3.2.1) translates into the assertion that,

$$\sum_{v \in S} \lambda_{D,v}(b) < (2 + \epsilon) h(b)$$

for almost all $b \in k$. But the above inequality remains unchanged if we replace k with a larger number field (cf. 3.4.1d); over $k(a_1, \ldots, a_n)$ this is now equivalent to (2.2.4). Thus this weakened version of the Second Main Theorem is equivalent to Roth's theorem.

By a similar argument, the Second Main Theorem itself translates into the assertion that,

$$(3.2.2) \qquad \prod_{v \in S} \min(1, \|b - \alpha_v\|_v) > \frac{1}{H(b)^2 h(b)^c}$$

for almost all $b \in k$; this has been conjectured by Lang [L 3], with $c = 1 + \epsilon$. He also has an analogue for abelian varieties [L 5, p. 783], [L 11].

§3. Defects

Returning to (3.2.1), we divide it by $T(r)$ and take limits. This gives the defect relation,

$$\sum_{a \in \mathbf{C}} \delta(a) \leq 2.$$

By a similar process, Roth's theorem implies,

$$\sum_{a \in k} \delta(a) \leq 2.$$

This defect relation is actually equivalent to Roth's theorem. Indeed, we have just shown one half of the equivalence; now assume the defect relation. Then we have $\sum_{i=1}^{n} \delta(a_i) \leq 2$ for distinct $a_1, \ldots, a_n$; by standard reductions appearing in the proof of Roth's theorem, any infinite sequence of points not satisfying (3.2.1) has a subsequence for which $\sum_{i=1}^{n} \delta(a_i) > 2$. See, for example, [**L 7**, Theorem 2.1, p. 163].

The defect δ is an interesting quantity which deserves a more detailed explanation. In the analytic case, the counting function N measures the growth of the number of zeroes of $f - a$. More generally, we will think of a as a divisor D and measure the area of f^*D. Usually (for most D) this will grow at the same rate as the characteristic function of f and, by the First Main Theorem, it will never grow faster. Establishing a lower bound for the size of f^*D is more subtle. The defect δ is defined to measure to what extent a divisor is deficient; i. e. f^*D does not grow as fast as possible. By the First Main Theorem, we have $0 \leq \delta \leq 1$ and if f never meets D then $\delta(D) = 1$. Lower bounds on the size of f^*D translate into upper bounds on δ. The estimate $\delta \leq 1$ is very crude since $\delta = 0$ for almost all D in a given linear system. This latter statement, in the one dimensional case, is a variant of the Liouville theorem:

$$\int \delta(a) = 0.$$

The defect relation is a much stronger version of this statement.

With a few changes in wording, all the above statements hold in the arithmetic case. Instead of measuring the growth of the area of f^*D, we measure the growth of the part of the denominators prime to S, in the map $V \setminus D \to \mathbf{P}^N$ determined by a basis for $\mathcal{L}(D)$. Usually this grows at the same rate as the multiplicative height, but never faster. The defect in this case measures how slowly the denominators grow; if the set $\mathcal{R}$ is D-integral, then the denominators are bounded and therefore $\delta(D) = 1$. This follows from Definition 1.4.3. If we assume that $a \in k$, then again we have $0 \leq \delta(a) \leq 1$. Since the major thrust of this note is the study of integral points, the most important application of defect relations will involve the question of finding out which divisors D can have $\delta(D) = 1$.

For example, consider Siegel's theorem, that there are only finitely many integral points on any affine variety isomorphic to $\mathbf{P}^1$ minus at least three points. This follows formally from the defect relation since, if D is the divisor on $\mathbf{P}^1$ consisting of those points, then each point $P \in D$ has $\delta(P) = 1$. Hence the sum

$$\sum_{P \in |D|} \delta(D) = \deg D > 2,$$

a contradiction.

The study of fractional values of δ can also be useful, though. A value $\delta(D) = 2/3$ means, for example, that the denominators described in the above paragraph grow at a rate of $H(b)^{1/3}$. The following example, based on ideas from Schmidt [**Schm 2**, § 7], gives a diophantine equation in which fractional values of the defect play a role.

Consider the norm form equation,

(3.3.1) $$N_{k/\mathbf{Q}}(a_1 X_1 + \cdots + a_n X_n) = c,$$

where k is a number field of degree d over $\mathbf{Q}$, $d > n$. Assume also that the factors are in general position; i. e. any n of them are linearly independent over $\mathbf{Q}$. Then, since no more than $n - 1$ of the factors, in absolute value, can be significantly smaller than $\max |X_i|$, each integral solution of (3.3.1) satisfies one of the finitely many inequalities,

(3.3.2) $$\prod |L_j(X)| \ll \max |X_i|^{-(d-n)}.$$

In the above expression, L_j is the jth conjugate of the linear form $a_1 X_1 + \cdots + a_n X_n$ and the product ranges over any n distinct values of j in the range $1 \leq j \leq d$. Applying Theorem 2.1.2 to (3.3.2) then implies that the solution set of (3.3.1) is degenerate.

The above discussion, however, does not use the full power of Theorem 2.1.2 since, in the latter, the exponent can be any negative number. Therefore, this slack can be absorbed by replacing the constant c in (3.3.1) with a polynomial in $X_1, \ldots, X_n$ of total degree less than $d - n$. Thus the equation,

$$(3.3.3) \qquad N_{k/\mathbf{Q}}(a_1 X_1 + \cdots + a_n X_n) = p(X)$$

has a degenerate set of solutions, provided that the degree of $p(X)$ is less than $d - n$.

We now rephrase the above argument in terms of the defect δ. The equation (3.3.1) defines a subset of $\mathbf{A}^n$ which can be mapped to $\mathbf{P}^{n-1}$ under the restriction of the canonical map $\mathbf{A}^n \setminus \{0\} \to \mathbf{P}^{n-1}$. This image omits the hyperplanes H_j defined by the factors L_j of (3.3.1). It can be shown that the images of solutions of (3.3.1) form a set whose defects $\delta(H_i)$ equal 1. Indeed, let $M_i(X)$ denote all monomials in $X_1, \ldots, X_n$ of degree d. Then the coordinates $M_i(X)/L_1(X) \ldots L_d(X)$ determine a map $(\mathbf{P}^{n-1} \setminus \bigcup H_j) \to \mathbf{A}^N$. Solutions of (3.3.1) mapped to $\mathbf{A}^N$ in this way have coordinates with bounded denominators since,

$$(3.3.4) \qquad \frac{M_i(X)}{L_1(X) \ldots L_d(X)} = \frac{M_i(X)}{c}.$$

Thus $\delta(\sum H_j) = 1$. This contradicts Schmidt's theorem in defect relation form (3.5.1.2), which states,

$$(3.3.5) \qquad \delta\left(\sum_{i=1}^d H_j\right) \leq \frac{(n-1) + 1}{d}.$$

Now an application of fractional defects can be demonstrated by applying the defect relation (3.3.5) to the general equation (3.3.3). In that case, by (3.3.4) (in which c is replaced by $p(X)$), the denominators grow at a rate of at most $H(X)^{\deg p}$. Relative to this embedding in $\mathbf{P}^N$, however, the height is equivalent to $H(X)^d$. Thus $\delta(\sum H_j) \geq 1 - (\deg p)/d$, which contradicts (3.3.5) if $\deg p < d - n$.

§4. The higher dimensional case

In this section we describe Nevanlinna theory in the higher dimensional case. This will require new definitions for the proximity, counting, and characteristic functions. After giving these definitions, we will be able to describe the Second Main Theorem and defect relations relative to a divisor on a variety. These can then be translated into a new conjecture which generalizes Roth's theorem to rational points on varieties.

The higher dimensional Second Main Theorem was due to Stoll [**St**]; here we follow Griffiths [**G**].

To begin the description of higher dimensional Nevanlinnna theory, let V be a nonsingular projective variety of dimension n. Nevanlinna theory studies holomorphic maps $f\colon \mathbf{C}^n \to V$. Instead of merely being non-constant, such maps are required to be non-degenerate; i. e. their Jacobian determinants do not vanish identically. This is slightly weaker than the notion of algebraic degeneracy, in which the image is not dense in the Zariski topology. (It is conjectured that the defect relations should also hold for maps that are algebraically non-degenerate.) In the number theoretic case we use the algebraic notion; i. e. an infinite set of distinct points $\mathcal{R} \subseteq V(k)$ which form a non-degenerate set as in Definition 1.4.13.

One of the major advances in Nevanlinna theory in recent years has been the introduction of differential-geometric methods. In particular, both the proximity function $m(a, r)$ and the characteristic function $T(r)$ have new definitions using metrics on line bundles. This corresponds well with the use of metrics in Arakelov theory.

The easiest definition to translate is the proximity function. Instead of using an element $a \in \mathbf{C}$, it uses a divisor D on V. On the analytic side, we let $|\cdot|$ be a metric on the associated line bundle $[D]$; also let

$$r^2 = |z_1|^2 + \cdots + |z_n|^2$$

so that r is the radius in $\mathbf{C}^n$; let $B[r]$ be the ball of radius r; and let σ be the rotationally invariant measure on the boundary $\partial B[r]$ such that

$$\int_{\partial B[r]} \sigma = 1.$$

Then $m(D, r)$ is defined by the equation,

$$m(D, r) = \int_{\partial B[r]} -\log |s(f(z))| d\sigma.$$

In the arithmetic context, we use summation over $v \in S$ instead of integration, and Weil functions $\lambda_{D,v}$ instead of hermitian metrics (cf. Section 1.3). The proximity function thus translates into,

$$m(D, P) = \sum_{v \in S} \lambda_{D,v}(P),$$

defined for $P \in V(k)$. This is the infinity component of the intersection pairing in the sense of Arakelov theory.

The counting function is defined as,

$$N(D,r) = \int_0^r \frac{dt}{t} \int_{f^*D \cap B[t]} \phi^{n-1}$$

where $\phi = dd^c r^2$; see [G]. If $f^*D = \sum n_i D_i$ for irreducible analytic divisors D_i, then we can rearrange this as before to give,

$$N(D,r) = \sum_i n_i \int_{D_i \cap B[r]} \phi^{n-1},$$

which translates into the expression,

$$N(D,P) = \sum_{v \notin S} \lambda_{D,v}(P).$$

Similarly, we define $N_1(r)$ as,

$$\int_0^r \frac{dt}{t} \int_{R \cap B[t]} \phi^{n-1}$$

where R is the ramification divisor of f. It is an analytic divisor on $\mathbf{C}^n$.

LEMMA 3.4.1. *The proximity and counting functions satisfy the following properties:*

(a). *If D and D' are divisors, then*

$$m(D + D', P) = m(D,P) + m(D',P) + O(1),$$
$$N(D + D', P) = N(D,P) + N(D',P) + O(1).$$

(b). *If D is effective, then $m(D,P)$ and $N(D,P)$ are bounded from below.*

(c). *Let $\phi: V \to W$ be a morphism of nonsingular complete varieties and let D be a divisor on W not contained in the image of ϕ. Then $m(\phi^*D, P) = m(D, \phi(P))$ and $N(\phi^*D, P) = N(D, \phi(P))$ (mod $O(1)$).*

(d). *$m(D,P)$ and $N(D,P)$ do not depend on the number field used in the definition. In other words, if E is a number field containing k and T is the set of places of E lying over places $v \in S$, then*

$$m_S(D,P) = m_T(D,P)$$

for all $P \in V(k)$, and likewise for $N(D,P)$.

PROOF: Follows from Lemma 1.3.3 and the definitions. □

It is now possible to define m and N on $V(\overline{k})$ by letting $E = k(P)$ and using part (b), above.

Finally, to describe the characteristic function, let L be a metrized line bundle, and let $\rho_i \colon U_i \to \mathbf{R}$ be the C^∞ functions defining the metric as in Section 1.3. The local expressions $dd^c \log \rho_i$ on U_i patch together, giving a global $(1,1)$ form which is the first Chern class $c_1(L)$. Then the characteristic function is,

$$T_L(r) = \int_0^r \left(\int_{B[t]} c_1(L) \wedge \phi^{n-1} \right) \frac{dt}{t^{2n-1}}.$$

This is automatically independent of linear equivalence, and also independent of the metric (mod $O(1)$). With some work, it is possible to show that the First Main Theorem holds. From this it follows that the assertions of Lemma 3.4.1 hold also for the characteristic function.

We say a real C^∞ $(1,1)$ form ϕ is <u>positive</u> (written $\phi > 0$) if locally

$$\phi = \sqrt{-1} \sum_{i,j} (\phi_{ij} dz_i \wedge d\bar{z}_j)$$

and (ϕ_{ij}) is a positive definite Hermitian matrix. A line bundle L is ample if and only if it has a metric ρ such that $c_1(\rho) > 0$. Due to a lack of references, we sketch the proof of this here. If L is ample, we can assume $L^{\otimes m} = f^* O(1)$ for some projective embedding f. Then, letting ρ be the pull-back of the Fubini-Study metric on $\mathbf{P}^N$, a little computation shows that $c_1(\rho) > 0$ on $L^{\otimes m}$, so $c_1(\rho^{1/m}) > 0$ on L. The converse is the Kodaira Embedding Theorem [G-H, p. 181]. Thus, as with heights, the characteristic function relative to an ample line bundle increases and goes to infinity as r grows.

The characteristic function does not have any obvious translation into number theory, however; therefore we define the height as before,

$$h_D(P) = \sum_v \lambda_{D,v}(P).$$

This definition automatically satisfies the First Main Theorem, but it takes a theorem to show that it is independent of linear equivalence.

Now if D is ample, then the defect is,

$$\delta(D) = \liminf \frac{m(D,r)}{T_D(r)}.$$

In the arithmetic case, we use the same expression with the height in the denominator, but we let the limit refer to Zariski-open subsets of V. In other words, $\delta(D)$ is the smallest real number A such that for all $\epsilon > 0$ and all non-empty Zariski-open subsets U of V, there is a point $P \in \mathcal{R} \cap U$ for which $m(D,P)/h_{[D]}(P) < A + \epsilon$.

By Lemma 3.4.1 (c) and (d), δ is functorial and does not depend on the number field. Also, as before, we have $0 \leq \delta \leq 1$.

The Second Main Theorem in the higher dimensional case is the following.

THEOREM 3.4.2 (STOLL, CARLSON-GRIFFITHS). *Let V be a nonsingular projective variety of dimension n. Let K be the canonical divisor of V, and let D be a normal crossings divisor on V (i. e. locally it is of the form $z_1 \ldots z_i = 0$ for some choice of analytic local coordinates $z_1, \ldots, z_n$). Then*

(a). *(Second Main Theorem) If A is an ample divisor on V and $\epsilon > 0$, then*

$$m(D,r) + T_K(r) < \epsilon T_A(r) - N_1(r). \qquad //$$

(b). *(Defect Relation) If D is ample, then*

$$\delta(D) \leq \liminf \frac{T_{-K}(r)}{T_D(r)}.$$

(Part (a) is usually stated with $A = [D]$, assumed to be ample. Our version follows from the usual version by applying the latter to an ample $D' > D$, with $A = [D']$ and then using Lemma 3.4.1 and Proposition 1.2.7.)

Since $N_1(r) \geq 0$, it can be ignored. We have included it here only for future reference (see Chapter 5).

It is possible to relax the above conditions somewhat, by letting V be a complete nonsingular variety and replacing ample divisors with almost ample divisors. Thus the corresponding statement in number theory is,

CONJECTURE 3.4.3 (MAIN CONJECTURE). *Let V be a nonsingular complete variety, defined over a number field k. Let D and K be as above, also defined over k. Then:*

(a). *If $\epsilon > 0$ and A is almost ample, then there exists a proper Zariski-closed subset $Z = Z(V, D, k, A, \epsilon, S)$ such that for all $P \in V(k)$,*

$$P \notin Z,$$
$$m(D,P) + h_K(P) \le \epsilon h_A(P) + O(1).$$

(b). If D is almost ample, then for all infinite Zariski-dense sets $\mathcal{R} \subseteq V(k)$,

$$\delta(D) \le \liminf \frac{h_{-K}(P)}{h_D(P)}$$
$$\le \inf\{p/q \in \mathbf{Q} \mid q > 0,\, pD + qK \text{ almost ample}\}.$$

The next section will contain an example to show why, in (a), the Zariski-closed subset Z is necessary. In (b) it appears implicitly in the Zariski limit. It is probably also true that the higher dimensional part of Z can be taken to depend only on V and D.

Also, we note that (b) is a special case of (a): divide (a) by $h_D(P)$ and take the lim inf. We include it here because it is sometimes more convenient for applications.

§5. Some Examples

We conclude this chapter with some examples to illustrate why various conditions in the conjecture are necessary. The first example shows how the theorems of Roth and Schmidt are special cases; it also gives an example in which the set Z is necessary. The next two examples indicate why it is necessary that D have normal crossings. The last example shows that the conjecture is well-behaved relative to blowups.

EXAMPLE 3.5.1. Let $V = \mathbf{P}^n$ and let $D = \sum H_i$ be a sum of hyperplanes in general position. Then the conjecture reduces to,

$$m(D,P) + h_K(P) < \epsilon h_{\mathcal{O}(1)}(P),$$

outside of a proper Zariski-closed subset. Since $K = \mathcal{O}(-n - 1)$, this becomes,

(3.5.1.1)
$$m(D,P) < (n + 1 + \epsilon)h(P),$$

which is the inequality of (2.2.4), Schmidt's Subspace Theorem. In defect relation form, this reads,

(3.5.1.2)
$$\delta(D) \le \frac{n + 1}{\deg D}.$$

Thus, the conjecture contains Schmidt's theorem, except for the statement that the Zariski-closed subset Z must consist of hyperplanes. In Schmidt's theorem, however, it is necessary that the exceptional set Z consist of hyperplanes, not points; therefore the Zariski-closed subset of (3.4.3) must allow higher dimensional components. On the other hand, as in Theorem 2.3.1, the higher dimensional components can be taken independently of the ring; this is the reason for conjecturing that this holds for Conjecture (3.4.3) as well.

The next two examples are derived from examples of Green [**Gr 2**]; they support the necessity of the assumption that components of D be smooth, meeting transversally.

EXAMPLE 3.5.2. Let $V = \mathbf{P}^2$ and let D be the divisor consisting of the lines $x = 0$ and $y = 0$ and the conic $(x - y)z = (x + y)^2$. This divisor has an ordinary triple point at $(0, 0, 1)$. Assume k is a number field having a unit ς which is not a root of unity. Then for arbitrary integers i, j, the point,

$$\left(\varsigma^i, 1, \varsigma^i + 3 - \frac{4(\varsigma^{ij} - 1)}{\varsigma^i - 1} \right)$$

is integral relative to D since x/y, z/y, y/x, z/x, and

$$\frac{y^2}{(x - y)z - (x + y)^2} = \frac{-1}{4\varsigma^{ij}},$$

etc. are integers. These points, however, form a Zariski-dense subset, hence the set is non-degenerate.

Geometrically, one sees that any line through $(0, 0, 1)$ not tangent to D will meet D in exactly one other point; therefore $\mathbf{P}^2 \setminus D$ contains infinitely many copies of $\mathbf{P}^1 - \{0, \infty\}$, which can have infinitely many integral points.

EXAMPLE 3.5.3. Again let $V = \mathbf{P}^2$ and let D be the divisor $yz^{d-1} = x^d$, $d \geq 4$. This has a bad singularity at $(0, 1, 0)$. If a and u are an integer and a unit, respectively, then the point,

$$(a, a^d + u, 1)$$

is integral relative to D. Such points again are dense in $\mathbf{P}^2$. Or, geometrically, one can cover $\mathbf{P}^2$ with lines $x = cz$ meeting D in only two points.

EXAMPLE 3.5.4. This last example concerns how the conjecture behaves relative to a blowup. Let V, K, D, A, and ϵ be as in Conjecture 3.4.3 and let B be a subvariety of codimension r. Let $\pi\colon V' \to V$ be the blowup of V along B, with exceptional divisor E. Assume furthermore that B is nonsingular and meets D transversally, so that V' and E will also be nonsingular. Moreover, $\pi^* D$ will have normal crossings, with the exception that E may occur in $\pi^* D$ with multiplicity $r' > 1$, where r' is the number of branches of D containing B. By our smoothness assumptions, $r' < r$, so that the conjecture on V' can give an upper bound on $m(\pi^* D - (r'-1)\dot{E})$. We also have $\pi^* A$ almost ample on V' and the canonical divisor of V' is $K' = \pi^* K + (r-1)E$. Thus the conjecture on V' asserts that,

$$m(\pi^* D - (r'-1)E) + h_{\pi^* K + (r-1)E} \le \epsilon h_{\pi^* A} + O(1)$$

on an open dense subset of $V'(k)$. This inequality reduces to,

$$m(\pi^* D) + h_{\pi^* K} + [(r-r')m(E) + (r-1)N(E)] \le \epsilon h_{\pi^* A} + O(1).$$

Except for the term in brackets, this coincides with the conjecture on V. The term in brackets is nonnegative, but very small, since the linear system $\mathcal{L}(nE)$ never has dimension larger than one. Thus, blowups, as restricted above, will never make the conjecture worse; in fact, they give a marginal improvement.

Chapter 4
Consequences of the Main Conjecture

In this chapter we obtain, as corollaries of the Main Conjecture (3.4.3), a number of diophantine conjectures that have already been made by others. In each case, the Main Conjecture gives an answer which is very close to what has already been conjectured.

In Section 1, we consider how the Main Conjecture applies to varieties of general type. It implies that sets of rational points cannot be Zariski-dense. This would give a positive answer to a question posed by Bombieri [N] for varieties of general type; also, it contains the Mordell conjecture.

In Sections 2 and 3 we show how the conjecture supports conjectures of Lang on integral points on abelian varieties and rational points on hyperbolic varieties, respectively. In the latter section, we discuss a variant of Nevanlinna theory for negatively curved varieties (see [G-K]). This gives a conjecture which contains the recently proved Shafarevich conjecture.

Finally, in Section 4 we derive Hall's conjecture, obtaining the same exponent as Hall had suggested.

§1. Degeneracy of Integral Points

Recall that a variety of general type is one whose canonical divisor is almost ample. If V is such a variety, then the Main Conjecture implies that $V(k)$ is degenerate for any number field k. Indeed, both the notion of general type and degeneracy of $V(k)$ are birational invariants, so we may assume V nonsingular. Then the Main Conjecture (with $D = 0$) implies that

$$h_K \leq \epsilon h_A + O(1)$$

on an open dense set. But taking $A = K$ and $\epsilon < 1$ implies that h_K is bounded, which is a contradiction unless $V(k)$ is degenerate.

Thus we have, conjecturally, an affirmative answer to the question posed by Bombieri [N], as to whether $V(k)$ is degenerate if V is a variety of general type. This also implies the Mordell conjecture, since curves are of general type if and only if their genus is at least two, and degeneracy on curves reduces to finiteness. Noguchi [N] proved this in the function field case, in the special case when the cotangent bundle is ample. This latter condition is strictly stronger than the condition of general type.

Using the following definition, we can make the same sort of statement for integral points.

DEFINITION 4.1.1 (IITAKA, [Ii, CH. 11]). *Let W be a nonsingular variety of dimension n, and assume $W = V \setminus D$, where D is a normal crossings divisor on a complete nonsingular variety V.*

(a) The <u>logarithmic canonical bundle</u> of W is the line bundle

$$\Omega^n[D] \cong K_V \otimes [D].$$

It is also isomorphic to the top exterior power of $\Omega^1[\log D]$, which is the vector bundle generated locally by the holomorphic differentials and the differentials $dz_1/z_1, \ldots, dz_i/z_i$, where locally D is given by $z_1 z_2 \ldots z_i = 0$.

(b) W is of <u>logarithmic general type</u> if its logarithmic canonical bundle is almost ample on V.

The above condition is independent of the representation $W = V \setminus D$.

PROPOSITION 4.1.2. *Let V and D be as in Definition 4.1.1, so that $V \setminus D$ is of logarithmic general type. Let $\mathcal{R} \subseteq V(k)$ be a set of D-integral points on V. Then $N(D, P)$ is bounded (Definition 1.4.3). If the Main Conjecture holds, then $\mathcal{R}$ is degenerate.*

PROOF: The conjecture implies that,

$$m(D, P) + h_K(P) \leq \epsilon\, h_A(P) + O(1)$$

outside of some proper Zariski-closed subset of V. Since $N(D, P)$ is bounded, the left-hand side reduces to h_{K+D}. Since $V \setminus D$ is of logarithmic general type, we may take $A = K + D$. Then, taking $\epsilon < 1$, we have $h_A(P)$ bounded as before, so $\mathcal{R}$ must be degenerate. □

Note that if $D = 0$, then the notion of integrality reduces to rationality, so that this proposition actually contains the above discussion on rational points.

In the case of curves, logarithmic general type covers exactly those curves for which finiteness holds by Mordell's conjecture and Siegel's theorems.

§2. Integral Points on Abelian Varieties

Let A denote an abelian variety and let D be an ample divisor on A. Then Lang has conjectured ([L 1]) that A has only finitely many D-integral points. In this section we show how this comes as a consequence of the Main Conjecture J. Silverman [Sil 3] has partially proved this, showing that given an abelian variety, there exists an ample divisor for which Lang's conjecture holds. See also Corollary 2.4.5.

LEMMA 4.2.1. *Let D be an ample divisor on an abelian variety A. Let V be a subvariety of A. Then (the desingularization of) $V \setminus D|_V$ is of logarithmic general type.*

PROOF: On an abelian variety A, the global holomorphic 1-forms generate the cotangent space at every point. Therefore the same is true for any blowup of A outside of the exceptional divisors of the blowup. Let $\pi: A' \to A$ be a composition of blowups of A such that $V' = \pi^{-1}(V)$ (the strict transform) is nonsingular. Then for almost all points of V, global holomorphic 1-forms again generate the cotangent space. Let $d = \dim V'$ and let $\theta_1, \ldots, \theta_d$ be global holomorphic 1-forms generating the cotangent space of V at some point P. Then $\theta_1 \wedge \cdots \wedge \theta_d$ is a section of the canonical divisor of V' which is holomorphic everywhere and nonzero at P. Thus the canonical divisor of V' is effective; since $\pi^* D|_{V'}$ is almost ample, so is $K_{V'} + \pi^* D|_{V'}$. Thus $V' \setminus \pi^* D|_{V'}$, and hence also $V \setminus D|_V$, is of logarithmic general type. $\qquad\square$

Now we apply the Main Conjecture to A and D. By (4.1.2) any set of integral points would be degenerate; we must show that the conjecture actually implies that all such sets must be finite. In other words, if Z is the Zariski closure of a set of integral points, then all irreducible components of Z must be zero dimensional. But if V is a higher dimensional component of Z, then $V \setminus D|_V$ must have a dense set of integral points, contradicting Lemma 4.2.1 and Proposition 4.1.2. Thus Lang's conjecture follows from the Main Conjecture.

§3. Lang's Conjectures on Curvature and Hyperbolicity

Lang conjectured, in [**L 5**], that a variety which is Kobayashi hyperbolic should have only finitely many rational points. Recently, in [**L 8**], he has added another conjecture that a variety has $V(k)$ degenerate (with fixed exceptional set in dimension > 0) if and only if the variety is measure hyperbolic. He also gives two separate lists of conditions on a projective variety defined over $\mathbf{C}$, and clearly summarizes the problem, due to many people, of whether the conditions in each list are equivalent. Statements in Nevanlinna theory support some of the implications, especially those dealing with curvature. Also see [**L 9**].

For Nevanlinna theory, the most relevant notion of curvature is the Griffiths function, which is defined as follows. Let ψ be an (n,n) form on the complete complex variety V, where $n = \dim V$. Then in local coordinates

$z_1, \ldots, z_n$ we have,

$$\psi = \rho(z) \prod_{i=1}^{n} \frac{\sqrt{-1}}{2\pi} dz_i \wedge d\bar{z}_i.$$

If ρ is a positive real C^∞ function, then we say ψ is a volume form. If we allow holomorphic zeroes, i. e. if $\rho = |f|^{2q}\sigma$ where f is holomorphic, σ is positive and C^∞, and $q > 0$ is rational, then ψ is called a pseudo volume form. In either case the functions $dd^c \log \rho = dd^c \log \sigma$ patch together to give a global $(1,1)$ form $\mathrm{Ric}(\psi) = c_1(\psi)$. Then the Griffiths function is the function $G(z)$ for which,

(4.3.1) $$\frac{1}{n!} \mathrm{Ric}(\psi)^n = G(z)\psi.$$

If $n = 1$ this coincides with minus the Gaussian curvature.

We also use the holomorphic sectional curvature K_s relative to a given positive $(1,1)$ form ω, as defined in [K]. This also reduces to the Gaussian curvature on curves; more generally it gives an upper bound on the curvature of one-dimensional analytic submanifolds. We say a variety is negatively curved if the sectional curvature is bounded from above by a negative constant.

Finally, Lang also refers to Kobayashi hyperbolicity. By definition, a manifold is Kobayashi hyperbolic if and only if the Kobayashi distance is positive; the Kobayashi distance is the largest distance agreeing with the Poincare distance on the unit disk, and such that every holomorphic mapping is distance-decreasing. For more details, see [K].

The first list of conditions is,

PROBLEM 4.3.2. *Let V be a projective algebraic variety. Determine which of the following conditions are equivalent:*

(1). *$V(k)$ is finite for any finitely generated extension field $k/\mathbf{Q}$;*

(2). *All subvarieties of V (including V itself) are of general type;*

(3). *V is negatively curved relative to some $(1,1)$ form ω;*

(4). *V is Kobayashi hyperbolic.*

(5). *Every subvariety of V is measure hyperbolic.*

Kobayashi has shown that (3) implies (4); otherwise all equivalences above remain unproved. He stated (4) $\implies$ (3) as a problem; other implications in the above list are conjectures of Lang. Before discussing how the Main Conjecture relates to the above problem, we note the "pseudofication" of the above list, namely:

PROBLEM 4.3.2P. *Let V be a projective algebraic variety. Determine which of the following are equivalent:*

(1p). *There exists a proper Zariski-closed subset Z of V, independent of k, such that $V(k) \setminus Z$ is finite for any field k of finite type over $\mathbf{Q}$;*

(2p). *V is of general type;*

(3p). *There exists a pseudo volume form ψ for which $\mathrm{Ric}\,\psi > 0$;*

(4p). *V is pseudo Kobayashi hyperbolic; i. e. there exists a proper Zariski-closed subset Z such that the Kobayashi distance satsfies $d(x,y) > 0$ unless $x = y$ or $x, y \in Z$.*

(5p). *V is measure hyperbolic (see [K]).*

Except for (3), the pseudo conditions are obviously weaker than their non-pseudo counterparts. Also, except for (5p), the pseudo conditions are known to be birational invariants. Currently what is known is that $(2p) \Longleftrightarrow (3p)$ ([**Tot**]), $(2p) \Longrightarrow (5p)$, $(4p) \Longrightarrow (5p)$, and $(5p) \Longrightarrow (2p)$ for surfaces (see [**M-M**]). Kobayashi posed $(5p) \Longrightarrow (2p)$ as a problem; other implications are conjectures of Lang.

In fact, Carlson and Griffiths first used $(2p) \Longrightarrow (3p)$ as a step in the proof of the Second Main Theorem of Nevanlinna theory, so the Main Conjecture actually supports $(3p) \Longrightarrow (1p)$ as well as $(2p) \Longrightarrow (1p)$.

As for the non-pseudo conjecture, first consider $(4) \Longleftrightarrow (1)$. This was Lang's original conjecture in this direction. This also works well with the dictionary of Table 3.1 in a slightly different way. Indeed, a variety V is hyperbolic if and only if, given a metric $|\cdot|$ on the tangent bundle, all holomorphic maps $f \colon \mathbf{D} \to V$ have $|df(0)|$ bounded. By Table 3.1, on the one hand, such maps correspond to rational points; by $(1) \Longleftrightarrow (4)$, on the other hand, $|df(0)|$ is bounded if and only if the height of the rational points is bounded.

The part $(2) \Longrightarrow (1)$ of this conjecture also follows from the Main Conjecture, as was the case with (2p).

For $(3) \Longrightarrow (1)$, Griffiths and King have used the methods of Nevanlinna theory to prove the following generalization of a theorem of Kwack:

DEFINITION 4.3.3. *Let ω be a $(1,1)$ form on a quasi-projective variety V. Then ω is <u>large</u> if there exists an ample line bundle L on V for which,*

(a). *There exists a metric ρ on L and a constant $A > 0$ such that,*

$$\omega > Ac_1(\rho);$$

(b). *There exist sections* $\sigma_0, \ldots, \sigma_N$ *of* L *defining a projective embedding of* V *into* $\mathbf{P}^N$; *and*

(c). *The lengths of the* σ_i *(relative to the metric* ρ) *are bounded.*

THEOREM 4.3.4 ([**G-K**, 9.20]). *Suppose* X *is a quasi-projective complex variety which is negatively curved with respect to a large* $(1,1)$ *form* ω. *Then any holomorphic map from an algebraic variety* A *(of any dimension) to* X *is rational.*

Kwack's theorem is the projective case of the above, for which all positive $(1,1)$ forms are large. This theorem supports the following conjecture.

CONJECTURE 4.3.5 (THE CONJECTURE USING $(1,1)$ FORMS). *Let* X *and* ω *be as above. Assume also that* $X = V \backslash D$, *where* V *is a complete variety and* D *is a divisor on* V. *Then any set of* D-*integral points on* V *is finite.*

Griffiths and King note that Theorem 4.3.4 applies also to quotients of bounded symmetric domains by arithmetic groups; hence Conjecture 4.3.5 also contains the recently proved Shafarevich conjecture. More generally, one might ask whether a version of Problems 4.3.2 and 4.3.2p should hold for integral points on non-compact varieties, subject to some additional conditions at infinity.

For a sketch of the proof of 4.3.4, as well as a further discussion of Conjecture 4.3.5 and abelian varieties, see also Section 5.7.

§4. Hall's Conjecture

The last consequence of the Main Conjecture that we will discuss is Hall's conjecture. In its original form (over $\mathbf{Z}$), it was posed as,

$$(4.4.1) \qquad |y^2 - x^3| \gg x^{\frac{1}{2} - \epsilon},$$

if x, y are integers with $y^2 \neq x^3$ [**Hal**]. Danilov [**Dan**] has proved that $1/2$ is the best possible exponent. We also have:

THEOREM 4.4.2 (MASON, SILVERMAN [**Sil 1**]). *Let* $k = k(C)$ *be the function field of a complete nonsingular curve* C, *let* S *be a finite set of places of* k, *and let* $\epsilon = 0$. *Then for all* S-*integers* $x, y \in k$,

$$h(y^2 - x^3) > (\tfrac{1}{6} - \epsilon)h([x^3 : y^2 : 1]) - O(1).$$

This latter inequality (with $\epsilon > 0$) seems to be the proper generalization of (4.4.1) to number fields. In this section we will show how it follows from applying the Main Conjecture to a certain divisor on a blowup of $\mathbf{P}^2$. It will also follow more easily from a conjecture in Chapter 5, but we will still demonstrate this in order to show the heuristics involved.

We start with (4.4.1); let $f(X, Y, Z) = Y^2/Z^2 - X^3/Z^3$ be a function on $\mathbf{P}^2$. The logarithm of the left-hand side of (4.4.1) can be written as $-m((f), P)$ where $P = [X : Y : Z]$ and m is extended to non-effective divisors by linearity. This is because $-\log|f|$ is a Weil function for the principal divisor (f). Therefore (4.4.1) can be expressed in terms of Nevanlinna theory as a conjecture that,

$$m((f), P) < (\tfrac{1}{2} - \epsilon)h_{\mathcal{O}(1)}(P) + O(1),$$

where P ranges over all integral points of $\mathbf{A}^2 \subseteq \mathbf{P}^2$.

We will want to apply the conjecture to the divisor $supp(f)$; however, that divisor is singular at infinity ($[0 : 1 : 0]$) and at the origin $[0 : 0 : 1]$. At the former point, the divisor $y^2 = x^3$ (call it D) meets the line $z = 0$ (call it Z) with multiplicity three. Therefore let $\pi_1 \colon V_1 \to \mathbf{P}^2$ be the blowup of this point. The preimage of this point is now a divisor on V_1; call it E_1. We have $E_1^2 = -1$, $K_{V_1} = \pi_1^* K_{\mathbf{P}^2} + E_1$, and E_1 has zero intersection with $\pi_1^* \operatorname{Pic}(\mathbf{P}^2)$ (cf. [H]).

In addition to pulling back a divisor via π_1^* (as a Cartier divisor), a divisor can also be pulled back as a strict transform: the strict transform of an irreducible divisor is the irreducible divisor on V_1 lying over the divisor on $\mathbf{P}^2$. Let $D^\#$ and $Z^\#$ be the strict transforms of D and Z, respectively. Since both D and Z are smooth at $[0 : 1 : 0]$, we have $D^\#.E_1 = Z^\#.E_1 = 1$, so that $\pi_1^* D = D^\# + E_1$, $\pi_1^* Z = Z^\# + E_1$. From the relation,

$$3 = (D.Z) = (\pi_1^* D . \pi_1^* Z) = (D^\# + E_1 . Z^\# + E_1)$$

and the known pairings with E_1, it follows that $(D^\#.Z^\#) = 2$; i.e. the intersection number drops by one.

Therefore it is necessary to blow up again, getting a map $\pi_2 \colon V_2 \to V_1$ with exceptional divisor E_2 and strict transforms which we will again denote $D^\#$, $Z^\#$, and $E_1^\#$. Again, the intersection number drops: $(D^\#.Z^\#) = 1$, but E_2 also passes through this point. Hence a third blowup $\pi_3 \colon V_3 \to V_2$ is necessary. Let E_3 be its exceptional divisor. We now have normal crossings

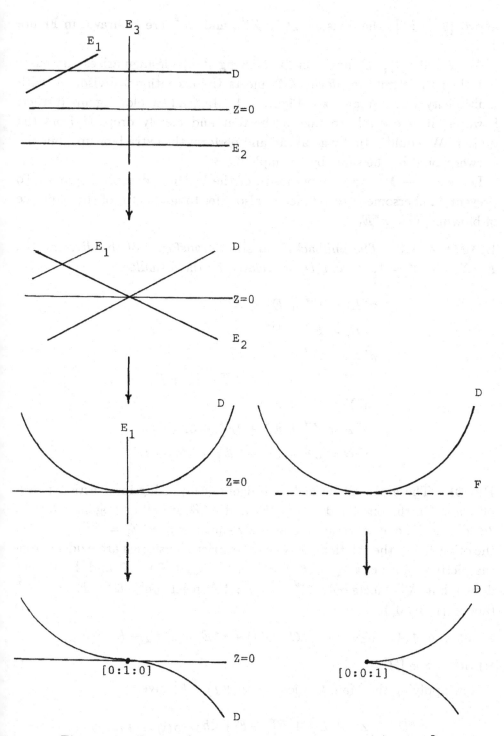

Figure 4.1. Desingularization of the support of (f) on $\mathbf{P}^2$.

above $[0:1:0]$; the divisors $D^{\#}$, $Z^{\#}$, and $E_i^{\#}$ are as drawn in Figure 4.1.

At $[0:0:1]$, D has a cusp; blowing it up desingularizes the cusp, but then the strict transform $D^{\#}$ meets the exceptional divisor F with multiplicity two. Again, see Figure 4.1. Instead of blowing up further, however, it is possible to take a shortcut and merely drop F from the divisor. We could actually go ahead and perform the extra blowups, but the answer would be the same by Example 3.5.4.

Let $\pi: V \to \mathbf{P}^2$ be the composite of the blowups described above. To prevent cumbersome notation, let π also refer to any factor of the sequence of blowups, viz. $\pi^* E_i$.

LEMMA 4.4.3. *The pullbacks and strict transforms of the divisors E_i, F, $X = 0$, $Y = 0$, Z and D are related by the formulas,*

$$\pi^* E_1 = E_1^{\#} + E_2^{\#} + E_3^{\#}$$
$$\pi^* E_2 = E_2^{\#} + E_3^{\#}$$
$$\pi^* E_3 = E_3^{\#}$$
$$\pi^* X = X^{\#} + E_1^{\#} + E_2^{\#} + E_3 + F$$
$$\pi^* Y = Y^{\#} + F$$
$$\pi^* Z = Z^{\#} + E_1^{\#} + 2E_2^{\#} + 3E_3$$
$$\pi^* D = D^{\#} + E_1^{\#} + 2E_2^{\#} + 3E_3 + 2F$$

PROOF: These can all be deduced algebraically using intersection multiplicities. On the one hand $\pi^* E_i$, F, and $\pi^* Z$ are all orthogonal; $3\pi^* X$, $3\pi^* Y$, $3\pi^* Z$, and $\pi^* D$ are numerically equivalent; $\pi^* E_i^2 = F^2 = -1$. On the other hand, the relations between the strict transforms are evident from the picture (since e. g. $(E_2^{\#}.E_3) = 1$, $(E_1^{\#}.E_3) = 0$). X and Y are not drawn, but $X^{\#}$ meets only $E_1^{\#}$ and F; $Y^{\#}$ meets only $D^{\#}$, F, and $Z^{\#}$ (above $[1:0:0]$). $\square$

LEMMA 4.4.4. $K_V = -3\pi^* \mathcal{C}al O(1) + \pi^* E_1 + \pi^* E_2 + E_3 + F$.

PROOF: See [H]. $\square$

Now applying the Main Conjecture to $|(f) - F|$ gives,

(4.4.5)
$$m(D^{\#} + Z^{\#} + E_1^{\#} + E_2^{\#} + E_3) < h_{3\pi^* O(1) - \pi^* E_1 - \pi^* E_2 - E_3 - F}$$
$$+ \epsilon h_A + O(1).$$

outside of some Zariski-closed subset. We want the left-hand side to be in the form,

$$m((f)) = m(D^{\#} - 3Z^{\#} - 2E_1^{\#} - 4E_2^{\#} - 6E_3 + 2F).$$

This leaves a discrepancy of $m(-4Z^{\#} - 3E_1^{\#} - 5E_2^{\#} - 7E_3 + 2F)$. But, since all points under consideration are integers (i. e. π^*Z-integers), we have $N(\pi^*Z, P) = O(1)$. Since N is additive in D and nonnegative, $N(4Z^{\#} + 3E_1^{\#} + 5E_2^{\#} + 7E_3, P) = O(1)$ (that divisor is contained in $4\pi^*Z$). Thus,

$$(4.4.6) \qquad -m(4Z^{\#} + 3E_1^{\#} + 5E_2^{\#} + 7E_3) = -h_{4Z^{\#} + 3E_1^{\#} + 5E_2^{\#} + 7E_3}.$$

Also, since $N \geq 0$,

$$(4.4.7) \qquad\qquad\qquad m(2F) \leq h_{2F}.$$

Adding (4.4.5), (4.4.6), and (4.4.7) gives,

$$(4.4.8) \qquad m((f)) \leq -h_{Z^{\#} + E_1^{\#} + E_2^{\#} + E_3 - F} + \epsilon\, h_A.$$

Letting L be the divisor $Z^{\#} + E_1^{\#} + E_2^{\#} + E_3 - F$, we can unravel the meaning of the height if we can compute $\mathcal{L}(L)$. We have,

$$\left(\frac{X}{Z}\right) = X^{\#} - Z^{\#} - E_2^{\#} - 2E_3 + F;$$

$$\left(\frac{Y}{Z}\right) = Y^{\#} - Z^{\#} - E_1^{\#} - 2E_2^{\#} - 3E_3 + F.$$

But from this it is not clear that $\mathcal{L}(nL)$ is ever nonempty; at least it never contains any power of X/Z or Y/Z. At first glance, there may be other elements of $\mathcal{L}(nL)$ for n large, but this is unlikely—since the exceptional curves have negative self intersection, it is probable that the only elements of $\mathcal{L}(nL)$ would be the above obvious ones.

The limiting factors in the above are the coefficients of E_3 and F. We are interested in the behavior at infinity, so the coefficient of E_3 is important. Therefore, we alter our expectations near the divisor F. Over $\mathbf{Z}$, this does not affect the conjecture, since integers are bounded away from 0. Hence, we consider estimates of $m((f)) - m(F)$ instead of $m((f))$. Replacing (4.4.7) with

$$(4.4.7b) \qquad\qquad\qquad m(F) \leq h_F$$

gives the inequality,

(4.4.8b) $$m((f)) - m(F) \leq -h_{Z^\# + E_1^\# + E_2^\# + E_3} + \epsilon\, h_A,$$

instead of (4.4.8). Letting $L' = Z^\# + E_1^\# + E_2^\# + E_3$, we see that $\mathcal{L}(6L')$ contains x^3, y^2, and 1; therefore, (4.4.8b) implies,

(4.4.9) $$m((f)) - m(F) \leq -\tfrac{1}{6} h([x^3 : y^2 : 1]) + \epsilon\, h_A.$$

Now for h_A we can actually use $h([x^3 : y^2 : 1])$. Indeed, we have a rational map $g: \mathbf{P}^2 \to \mathbf{P}^2$, defined by $[x : y : 1] \mapsto [x^3 : y^2 : 1]$. Pulling this up by π gives a map $g: V \to \mathbf{P}^2$ which is now a morphism. Hence $g^*\mathcal{O}(1)$ is almost ample (Lemma 1.2.6e), and $h_{g^*\mathcal{O}(1)} = h([x^3 : y^2 : 1])$.

Finally, since the left-hand side of (4.4.9) is,

$$\sum_{v \in S} \left(-\log \|y^2 - x^3\|_v + \log \frac{\max(\|x\|_v, \|y\|_v)}{\max(\|x\|_v, \|y\|_v, 1)} \right)$$
$$> \sum_{v \in S} -\log \max(\|y^2 - x^3\|_v, 1),$$

(4.4.9) is actually a refinement of (4.4.2).

In the above, we have ignored the condition of the exceptional set. In this particular case, it is possible to eliminate it by a trick (although in general finding an exhaustive list of such subvarieties is an open problem).

In our situation, we have $\mathbf{G}_m$ acting on V by

$$u \cdot (x, y) = (u^2 x, u^3 y).$$

Let $C_1, \ldots, C_n$ be curves comprising the higher dimensional part of the exceptional closed subset. It is possible to show that each $u \in \mathbf{G}_m$ must permute these C_i's; since the set of u's is infinite, these curves must be fixed by $\mathbf{G}_m$. But then they must be of the form $y^2 = ax^3$; a simple computation then shows that all such curves satisfy the conjecture.

Chapter 5
The Ramification Term

In this chapter we introduce a generalization of the main conjecture of Chapter 3. This generalization extends the conjecture to all points P of a variety for which $[k(P) : \mathbf{Q}]$ is bounded. It also provides an interpretation for the ramification term $N_1(r)$, which has been ignored so far.

The basis for this conjecture is a slight generalization of the Chevalley-Weil theorem which will be proved in Section 1. If $\pi: V \to W$ is a generically finite (possibly ramified) map, then this generalization allows us to bound the discriminant of $k(\pi^{-1}(P))$ as P varies over the k-rational points of W.

In Section 2 we consider how this relates to the Main Conjecture of Chapter 3. If, as in Example 3.5.4, the statement of the conjecture remains largely unchanged by lifting, then the analogue of the ramification term in Nevanlinna theory should be an expression depending on the discriminant of $k(P)$. Thus this generalization is not as well supported as the main conjecture; in particular, not even the case for $\mathbf{P}^1$ has been proved.

The ramaining sections are largely independent of one another. In Section 3 we consider the special case over $\mathbf{P}^1$ (the "Roth" case). Section 4 contains other shorter remarks on the new conjecture. In Section 5 we show how the conjecture implies the "abc" conjecture and some of its corollaries; these include the "asymptotic Fermat" conjecture and a simpler derivation of the Hall conjecture. In Section 6, we prove the conjecture for curves over function fields in the split case, and in Section 7 we add a ramification term to the "Conjecture with $(1,1)$ forms" of Chapter 4.

§1. A Generalized Chevalley-Weil Theorem

Before stating the theorem, let us digress briefly on the topic of discriminants. If L/k is an extension of number fields, then

$$(5.1.1) \qquad D_{L/\mathbf{Q}} = D_{k/\mathbf{Q}}^{[L:k]} N_{\mathbf{Q}}^k D_{L/k}.$$

This follows from multiplicativity of the different in towers. If we introduce the absolute (logarithmic) discriminant,

$$(5.1.2) \qquad d(k) = \frac{1}{[k : \mathbf{Q}]} \log D_{k/\mathbf{Q}}$$

then (5.1.1) becomes,

$$d(L) - d(k) = \frac{1}{[L:Q]} \log N_{\mathbf{Q}}^k D_{L/k}$$

(5.1.3)
$$= \frac{1}{[L:Q]} \sum_{v \in M_k \setminus S_\infty} -\log \|D_{L/k}\|_v$$

REMARK 5.1.4. For any fixed rational prime p, the part of $d(k)$ coming from primes above p is bounded by $(1 + \log_p(n)) \log p$ (the second term comes from wild ramification; $n = [k:Q]$). In fact, this even applies to the part of $d(L) - d(k)$ attributable to primes over p if $n = [L:k]$.

Also, we note that, in the case of number fields, the different is the same as the annihilator of the sheaf of relative differentials Ω:

$$D_{L/k} = \operatorname{Ann} \Omega_{O_L/O_k};$$

thus

(5.1.5) $$d(L) - d(k) = \frac{1}{[L:Q]} \sum_{w \in M_L - S_\infty} (f_w \log p) \operatorname{length}_w \Omega_{O_L/O_k}.$$

Returning to varieties over k, we recall that the <u>ramification divisor</u> $R = R_{V/W}$ of a generically finite map of normal varieties $f: V \to W$, is defined as,

$$R = \sum_C \operatorname{length}_C \Omega_{V/W} \cdot C$$

where the sum extends over all prime divisors C of V. Here length_C means the length of the localized module at C. "Generically finite" means finite on an open dense set of points.

By abuse of notation, let $d(P) = d(k(P))$ if P is a closed point of a variety. Also, we recall from Chapter 3 that $m(D, P)$ and $N(D, P)$ are defined over the set of <u>algebraic</u> points on V, by Lemma 3.4.1(d).

THEOREM 5.1.6. Let $\pi: V \to W$ be a generically finite separable surjective map of complete nonsingular varieties, with ramification divisor R. Then for all separable algebraic points $Q \in V \setminus R$ and $P = \pi(Q)$, we have,

$$d(Q) - d(P) \leq N(R, Q) + O(1).$$

PROOF: Both sides split up into sums over places not in S (cf. 5.1.5); the general strategy will be to prove the inequality termwise for almost all places not in S. For the finitely many places remaining, the left-hand side is bounded, by Remark 5.1.4, and can be absorbed into the $O(1)$ term. We may also assume that V, W, and π are all defined over a given number field k. Let X and Y be schemes over $\mathcal{O}_k$ whose generic fibres (over k) are V and W respectively. Let π also denote the map $X \to Y$ corresponding to $V \xrightarrow{\pi} W$. Since V and W are nonsingular, X and Y have good reduction almost everywhere; therefore, enlarging S, we may assume that X and Y are smooth over primes outside S.

Points P and Q (defined over K and L, resp.) correspond to sections,

$$s\colon \operatorname{Spec} \mathcal{O}_K \to Y \quad \text{and} \quad t\colon \operatorname{Spec} \mathcal{O}_L \to X.$$

The definition of the ramification divisor generalizes directly to schemes, giving a divisor $R_{X/Y}$ on X whose generic fibre over k is $R_{V/W}$. As in the proof of Proposition 1.4.7, we have,

$$(5.1.7) \qquad N_S(R, Q) = \frac{1}{[L : \mathbf{Q}]} \sum_{\substack{w \in M_L \\ w \mid v \\ v \notin S}} (f_w \log p)\, \mathrm{length}_{\mathcal{O}_w}(\mathcal{O}_w/(g(t(w)))),$$

where g locally defines the divisor $R_{X/Y}$. We want to show that this dominates (5.1.5). As noted above, we will prove it termwise; therefore, it will suffice to prove the following lemma.

LEMMA 5.1.8. *Let $\mathcal{O}_v$ be a discrete valuation ring, let X and Y be smooth proper schemes over $\operatorname{Spec}\mathcal{O}_v$ of relative dimension n, and let $\pi\colon X \to Y$ be a generically finite surjective morphism. Let $s\colon \operatorname{Spec}\mathcal{O}_v \to Y$ be a section corresponding to a point $P \in W$, and let $t\colon \operatorname{Spec}\mathcal{O}_w \to X$ be a morphism lying over s; in other words, using the notation of the proof of the Theorem, w is a place of $k(Q)$ lying over v, $\pi'\colon \operatorname{Spec}\mathcal{O}_w \to \operatorname{Spec}\mathcal{O}_v$ corresponds to the inclusion $\mathcal{O}_v \subseteq \mathcal{O}_w$, and $\pi \circ t = s \circ \pi'$. Assume that $\Omega_{X/Y}$ is not supported on the generic point of the image of t, and that $\mathcal{O}_w$ is separable over $\mathcal{O}_v$.*

$$
\begin{array}{ccc}
\operatorname{Spec}\mathcal{O}_w & \xrightarrow{\ t\ } & X \\
\big\downarrow{\scriptstyle \pi'} & & \big\downarrow{\scriptstyle \pi} \\
\operatorname{Spec}\mathcal{O}_v & \xrightarrow{\ s\ } & Y
\end{array}
$$

(a). *The "First Exact Sequence",*

$$(5.1.9) \qquad \pi^*\Omega_{Y/\operatorname{Spec}\mathcal{O}_v} \to \Omega_{X/\operatorname{Spec}\mathcal{O}_v} \to \Omega_{X/Y} \to 0$$

is exact on the left also, and the first two terms are locally free of rank n.

(b). *Locally choose bases for the first two terms of (5.1.9) on an open set U. Let the function g_U be defined as the determinant of the first map in (5.1.9), relative to these bases. Then the ramification divisor R_π is locally given by (g) on U.*

(c). *Let U and g be as defined in part (b), and assume U contains the image of t. Then*

$$\text{length}_{O_w}(O_w/(g(Q))) = \text{length}_{O_w} t^*\Omega_{X/Y}.$$

(d).

$$\text{length}\, \Omega_{O_w/O_v} \leq \text{length}\, t^*\Omega_{X/Y}.$$

PROOF: (a) The second assertion follows from the assumption that X and Y are smooth over $\text{Spec}\, O_v$. The first assertion follows from the fact that $\Omega_{X/Y}$ is a torsion sheaf; therefore, the first map must be of full rank. The kernel is then a rank zero submodule of a free module; i. e. 0.

(b) We note that changing the choices of bases only changes g by a unit in $O(U)$; therefore (g) is a well-defined Cartier divisor. Let C be any prime divisor meeting U; then the result follows immediately, after localizing at C, because any matrix over a PID is equivalent to a diagonal matrix.

(c) Since the generic point of the image of t does not lie in the support of $\Omega_{X/Y}$, the conclusions of (a) apply also to the sequence,

$$0 \rightarrow (\pi \circ t)^*\Omega_{Y/\text{Spec}\, O_v} \rightarrow t^*\Omega_{X/\text{Spec}\, O_v} \rightarrow t^*\Omega_{X/Y} \rightarrow 0.$$

Choosing bases compatible with those chosen in (b) (via t^*), we find that the determinant of the first map in this case is $g(t(\eta)) = g(Q)$, where η is the generic point of $\text{Spec}\, O_w$. The conclusion is then immediate.

(d) Let $T = X \times_Y \text{Spec}\, O_v$ via s. Let $t': T \rightarrow X$ be the first projection and let $t'': \text{Spec}\, O_w \rightarrow T$ be the unique map such that $t = t' \circ t''$.

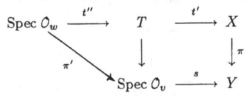

Then $t^*\Omega_{X/Y} \cong t''^*\Omega_{T/\text{Spec}\, O_v}$ by the properties of Ω relative to base change. Also, t'' is a closed immersion; otherwise O_w would be too large. Thus, by the "Second Exact Sequence",

$$t''^*\Omega_{T/\text{Spec}\, O_v} \twoheadrightarrow \Omega_{O_w/O_v}.$$

This gives the desired inequality of lengths. $\qquad\qquad\qquad\qquad$ □

Although it will not be used in the sequel, we include the following lemma for the sake of completeness:

LEMMA 5.1.10. *Let A be an almost ample divisor on a variety V. Then there exists a proper Zariski-closed subset $Z \subset V$ such that for all algebraic points $P \in V \setminus Z$ of bounded degree over $\mathbf{Q}$,*

$$d(P) = O(h_A(P))$$

for $h_A(P)$ sufficiently large.

PROOF: Let $\Phi: V \to \mathbf{P}^N$ be a birational map determined by some linear system $\mathcal{L}(nA)$, for some sufficiently large integer n. Then it will suffice to show that the lemma holds on $\mathbf{P}^N$ relative to $\mathcal{O}(1)$, provided that Z consists only of linear subspaces. But $\mathbf{P}^N$ is birationally equivalent to $(\mathbf{P}^1)^N$, so that we can reduce to the $\mathbf{P}^1$ case. Indeed, the rational map

$$[x_0 : x_1 : \cdots : x_n] \mapsto ([x_0 : x_1], [x_0 : x_2], \ldots, [x_0 : x_n])$$

is undefined only on certain linear subspaces, and by Kodaira's lemma (1.2.7), we may replace $\mathcal{O}(1)$ on $\mathbf{P}^N$ with $\mathcal{O}(1) \times \cdots \times \mathcal{O}(1)$ on $(\mathbf{P}^1)^N$. Relative to the latter line bundle, the height is just the sum of the heights of the coordinates:

$$h_{\mathcal{O}(1) \times \cdots \times \mathcal{O}(1)}(P_1, \ldots, P_N) = h(P_1) + \cdots + h(P_N) + O(1).$$

Also,

$$d((P_1, \ldots, P_N)) \leq d(P_1) + \cdots + d(P_N).$$

This reduces the problem to the case on $\mathbf{P}^1$ (or $\mathbf{A}^1$); in that case it is just elementary number theory. See Section 5.3, up through (5.3.2). $\qquad$ □

§2. The Ramification Term

Let $\pi: V \to W$ be a generically finite surjective morphism of complete nonsingular varieties. Let D be a reduced divisor on W. By sufficiently blowing up W and replacing D with the support of its pull-back, we may assume that W is nonsingular and that D has normal crossings. Replacing V with its normalization over the new W, and blowing it up sufficiently, we may assume that V is nonsingular and that the support of π^*D has normal crossings.

For the remainder of this section, assume that $\pi: V \to W$ is a generically finite surjective morphism of complete nonsingular varieties of dimension n. Also assume that D (resp. $|\pi^*D|$) are normal crossings divisors on W (resp. V).

In this situation, we will want to compare the Main Conjecture applied on W with the same conjecture on V.

LEMMA 5.2.1. *Let the canonical divisors on V and W be denoted K_V and K_W. Then*

$$K_V = \pi^*K_W + R.$$

PROOF: [**Mum**, 6.20]. This also follows from applying the formalisms of the determinant operation in K-theory (e. g. [**Man**]) to the exact sequence (5.1.9). □

LEMMA 5.2.2. $|\pi^*D| \geq \pi^*D - R$.

PROOF: The question is local on X, so let $P \in W$ and $Q \in V$ be closed points with $P = \pi(Q)$; let $y_1, \ldots, y_n$ (resp. $x_1, \ldots, x_n$) be local analytic coordinates for neighborhoods of P (resp. Q) on W (resp. V). Assume, moreover, that D is locally given by the equation $y_1 \ldots y_s = 0$, and that $|\pi^*D|$ is locally given by $x_1 \ldots x_r = 0$. Then each y_j, $1 \leq j \leq s$, can be expressed in terms of the x_i as,

$$y_j = \prod_{i=1}^{r} x_i^{a_{ij}} \cdot u_j(x_1, \ldots, x_n)$$

where u_j is a unit in $\mathbf{C}[[x_i, \ldots, x_n]]$.

Now, if one computes $dy_1 \wedge \cdots \wedge dy_n$ in terms of $dx_1 \wedge \cdots \wedge dx_n$, one finds that,

(5.2.3) $\qquad dy_1 \wedge \cdots \wedge dy_n = p \cdot dx_1 \wedge \cdots \wedge dx_n$

and that $p \in \mathbf{C}[[x_1, \ldots, x_n]]$ is divisible by $\prod_i x_i^{\Sigma_j a_{ij} - 1}$. Indeed, if one were to write out the computation leading to (5.2.3) and erase all the letters "d", then the result would be divisible by $\prod_i x_i^{\Sigma_j a_{ij}}$.

But R is locally given by the equation $p = 0$; therefore,

$$R \geq \sum_{i=1}^{r}\left(\sum_{j=1}^{s} a_{ij} - 1\right) \cdot (x_i)$$
$$= \pi^*D - |\pi^*D|.$$
□

Now we consider how the conjecture on W relates to the conjecture on V. For the present discussion we will ignore the exceptional Zariski-closed subset Z. On V, one would like to apply the conjecture to π^*D; however, it has multiple components where π is ramified. We may consider only $|\pi^*D|$. Now, ignoring for a moment the fact that points Q lying above $P \in W(k)$ may not be defined over any particular number field, the conjecture reads,

$$(5.2.4) \qquad m(|\pi^*D|, Q) + h_{K_V}(Q) \leq \epsilon h_A(Q) + O(1),$$

where A is as usual an almost ample divisor on V. Applying Lemmas 5.2.1 and 5.2.2, this implies,

$$(5.2.5) \qquad m(\pi^*D, Q) + h_{\pi^*K_W}(Q) + N(R, Q) \leq \epsilon h_A(Q) + O(1).$$

Since $d(P)$ is bounded for $P \epsilon W(k)$, we have by Theorem 5.1.6,

$$d(Q) \leq N(R, Q) + O(1)$$

for all $Q \in \pi^{-1}(V(k))$. Thus, if instead of (5.2.4) we had proposed that,

$$m(\pi^*D - R, Q) + h_{K_V}(Q) - d(Q) \leq \epsilon h_A(Q) + O(1),$$

then (5.2.5) would have read,

$$m(\pi^*D, Q) + h_{\pi^*K_W}(Q) + (N(R, Q) - d(Q)) \leq \epsilon h_A(Q) + O(1)$$

for all points $Q \in \pi^*W(k)$, which would yield a marginal improvement in the conjecture on $W(k)$. Therefore, as in Example 3.5.4, the form of such a conjecture would remain largely unchanged after lifting by a (possibly ramified) generically finite cover. For this reason, we conjecture the following.

CONJECTURE 5.2.6 (THE GENERAL CONJECTURE). *Let V be a complete nonsingular variety with canonical divisor K. Let D be a normal crossings divisor on V, and let k be a number field over which V and D are defined. Let A be an almost ample divisor on V. Let d be as in (5.1.2). Finally, let $\epsilon > 0$. Then*

(a). *If $\pi: V \to W$ is a finite map to a nonsingular complete variety W, then there exists a Zariski-closed subset $Z = Z(\pi, V, D, k, S, \epsilon, A)$ such that,*

$$m(D, P) + h_K(P) \leq d(P) + \epsilon h_A(P) + O(1)$$

for all points $P \in V \setminus Z$ for which $\pi(P) \in W(k)$.

(b). If r is a positive integer then there exists a Zariski-closed subset $Z = Z(r, V, D, k, S, \epsilon, A)$ such that,

$$m(D, P) + h_K(P) \leq (\dim V)d(P) + \epsilon h_A(P) + O(1)$$

for all points $P \in V \setminus Z$ for which $[k(P) : k] \leq r$.

For curves, (b) contains (a); otherwise it is neither weaker nor stronger. The factor $\dim V$ has not been discussed above, but it is probably necessary in light of Remark 5.4.2 and Example 5.7.6, below. In short, $d(P)$ should play the role of the ramification term $-N_1(r)$ (or $N(B, r) - N(R, r)$ in [G-K]) in Nevanlinna theory. See also Remark 6.5.12.

In order to combine parts (a) and (b), one would need a new "discriminant" $d'(P)$ satisfying,

$$d(k(P)) \leq d'(P) \leq (\dim V)d(k(P)) \quad (\mathrm{mod} \ O(1))$$

and the Chevalley-Weil formalisms of the previous section. One possibility might be to assign coordinates x_i to each of a finite collection of open sets covering V and take $d'(P) = \sum d(k(x_i(P)))$. However, we prefer the above more intrinsic formulation.

As was the case in Chapter 3, the higher dimensional part of Z should be independent of k, S, ϵ, and A. Also, it is plausible that this conjecture holds over $\overline{\mathbf{Q}}$; i. e. $r = \infty$. It is also plausible that the error term could be refined to $c \log h_A(P)$ as in (3.2.2).

Szpiro [Sz 1] has proved an effective bound for the heights of rational points on a curve of genus $g \geq 2$ defined over a function field of genus g. If E is a section of a semi-stable non-isotrivial fibration $f: X \to C$ with s points of bad reduction and degree of inseparability e, then

$$-(E.E) \leq 8p^e 3^{3g+1}(g-1)^2 \left(s + 1 + \frac{2q-2}{3^g} + \frac{1}{3^{3g}} \right).$$

The number $-(E.E)$ is related to the height h_K by the adjuction formula; also $2q - 2$ is equivalent to $d(P)$. Thus he has a bound which is linear in $d(P)$, but unfortunately the coefficient is greater than $1 + \epsilon$. He also proposed a diophantine conjecture in [Sz 2].

§3. The $\mathbf{P}^1$ Case of the General Conjecture

Let us examine this conjecture in the special case when $V = \mathbf{P}^1$. The conjecture would then imply a generalization of Roth's theorem in which the elements β could vary over all numbers of bounded degree over $\mathbf{Q}$, and the main inequality would read,

$$(5.3.1) \qquad \prod_{v \in S} \|\beta - \alpha_v\|_v < \frac{c}{H(\beta)^{2+\epsilon} D(\beta)},$$

where $D(\beta)$ is the discriminant (in the usual sense) of $\mathbf{Q}(\beta)$.

Even though this is the simplest possible case of the conjecture, no such bounds have been proved that use the discriminant. Wirsing [**W**] has proved a similar statement in which the inequality is,

$$\prod_{v \in S} \|\beta - \alpha_v\|_v < \frac{c}{H(\beta)^{2r+\epsilon}}$$

and β is of degree r over $\mathbf{Q}$. Since $D(\beta) \ll H(\beta)^{2r-2}$, this would follow from (5.3.1). Schmidt ([**Schm 1**, p. 278]) has proved the same inequality with exponent $r + 1 + \epsilon$, provided that S contains only one infinite place and α_v has degree greater than r. He does this by applying his Subspace Theorem to the coefficients of the minimal polynomial of β over $\mathbf{Q}$. The requirement on the degree of α is needed to imply that none of the exceptional subspaces contains a point rational over $\mathbf{Q}$.

We now examine how Schmidt's argument applies to (5.3.1). Let β be an algebraic number of degree at most r; without loss of generality we may assume that the degree is exactly r. Let $f(X)$ be its minimal polynomial over $\mathbf{Q}$; let a_r be the least common multiple of the denominators of the coefficients of f, so that $a_r f(X) = a_r X^r + \cdots + a_0$ is a polynomial with relatively prime integer coefficients. We let it correspond to a point $f \in \mathbf{P}^r$. A definition of the height which preceded Weil's definition is that $H(\beta) = \max_i |a_i|$; over all algebraic numbers of fixed degree, it is equivalent to the (multiplicative, relative) Weil height. Therefore, using absolute heights, we have,

$$(5.3.2) \qquad h(\beta) = \tfrac{1}{r} h_{\mathbf{P}^r}(f).$$

Now we consider the proximity term. For simplicity, assume that $D = (\alpha_1) + \cdots + (\alpha_m)$ where α_i are rational and finite, and that $S = S_\infty$.

If β is close to α, then $f(\alpha)$ should be small, so that f is close to the hyperplane $H: \alpha^r X_r + \cdots + X_0 = 0$. At finite places,

$$\prod_{v \notin S_\infty} \max(1, \|\beta\|_v) = \|a_r\|;$$

at infinity,

$$\prod_{v \in S_\infty} \min(1, \|\beta - \alpha\|_v) \max(1, \|\beta\|_v) \gg\ll \prod_{v \in S_\infty} \|\beta - \alpha\|_v.$$
$$= |f(\alpha)|$$

Combining these gives,

$$r(m(\alpha, \beta) + h(\beta)) = m(H, f) + h(f) + O(1);$$
(5.3.3) $$\qquad rm(\alpha, \beta) = m(H, f) + O(1).$$

Hence, let $H_1, \ldots, H_m$ be the hyperplanes associated to $\alpha_1, \ldots, \alpha_m$. They are in general position by the van der Monde determinant. Applying the defect relation version of Schmidt's subspace theorem (3.5.1.1) gives, as in [**Schm 1**],

$$\sum m(\alpha, \beta) \leq (r + 1 + \epsilon)h(\beta) + O(1).$$

To refine this further, we need to consider the discriminant of the polynomial f. This is given by a polynomial in $a_0/a_r, \ldots, a_{r-1}/a_r$. Multiplying by a_r^{2r-2} turns it into a homogeneous polynomial g of degree $2r - 2$. Then the discriminant of the field generated by a root of f will be at least the square free part of $g(f) = g(a_0, \ldots, a_r)$. The latter is, up to a factor of four, the discriminant D of the field $\mathbf{Q}(\sqrt{g(f)})$.

Let $\pi: V \to \mathbf{P}^r$ denote the normalization of $\mathbf{P}^r$ in the function field $k(\mathbf{P}^r)(\sqrt{g})$. It is ramified exactly above the divisor $g = 0$. We will apply part (a) of the General Conjecture to the divisor $\sum \pi^* H_i$, ignoring for now any questions of smoothness. We have,

$$m\left(\sum \pi^*(H_i), Q\right) + h_{K_V}(Q) \leq \tfrac{1}{2} \log D + \epsilon h_A(Q)$$

for points $Q \in V$ lying above f. By Lemma 5.2.1 this implies that on $\mathbf{P}^n$,

$$m\left(\sum H_i, f\right) \leq (r + 1)h_{O(1)}(f) - h_R(\pi^{-1}(f)) + \tfrac{1}{2} \log D + \epsilon h_A(f).$$

Since $R = \frac{1}{2}\pi^* D \sim \pi^* \mathcal{O}(r-1)$, the two height terms combine to give $2h_{\mathcal{O}(1)}$; then applying (5.3.2) and (5.3.3), we have on the original projective line the inequality,

$$\sum m(\alpha_i, \beta) \le (2+\epsilon)h(\beta) + \tfrac{1}{2}d(\beta),$$

outside of some subset of $\mathbf{P}^1$ corresponding to a Zariski-closed subset of $\mathbf{P}^r$.

The factor $1/2$ appearing in front of $d(\beta)$ would appear to indicate that this actually implies that (5.3.1) would hold with an exponent of $1+\epsilon$; however, one must also consider the exceptional Zariski-closed subset. Let us examine the simplest possible case, namely a line $L \subseteq \mathbf{P}^r$. Let $j: \mathbf{P}^1 \to \mathbf{P}^d$ be an injection whose image is L. Also assume $L \not\subseteq H_\alpha$ for any α; then $j^* H_\alpha$ consists of exactly one point. Let C be a curve which is isomorphic to the desingularization of the preimage of L on V. By abuse of notation, we denote the map $C \to \mathbf{P}^1$ again by π. Then $\pi: C \to \mathbf{P}^1$ is ramified only above $j^* D$; in particular, $\deg R \le 2r - 2$. Let us assume that $\deg R = 2r - 2$. Now $m(H_{\alpha_i})$ on $\mathbf{P}^r$ pulls back to $m(j^* H_{\alpha_i})$ on $\mathbf{P}^1$. Since the H_{α_i} are in general position, each point $x \in \mathbf{P}^1$ can be contained in at most r of the $j^* H_{\alpha_i}$, so

$$\sum_i m(j^* H_{\alpha_i}, x) \le r \sum_k m_{\mathbf{P}^1}(a_k, x) \le (2r + \epsilon)h(x) + O(1)$$

for $x \in \mathbf{P}^1$. On C, we have, conjecturally,

$$\sum_i m(\pi^* j^* H_{\alpha_i}, \pi^{-1}(x)) + r\, h_{-4+R}(\pi^{-1}(x))$$

$$\le r\, d(\pi^{-1}(x)) + \epsilon\, h(\pi^{-1}(x)) + O(1).$$

Since $d(\pi^{-1}(x)) \le N(R, \pi^{-1}(x)) \le h_R(\pi^{-1}(x))$, cancelling $(r-2)d(\pi^{-1}(x))$ gives,

$$\sum_i m(\pi^* j^* H_{\alpha_i}, \pi^{-1}(x)) \le (2+\epsilon)h_{\pi^* \mathcal{O}(1)}(\pi^{-1}(x)) + \log D + O(1);$$

On the original projective line,

$$\sum_i m(\alpha_i, \beta) \le (2+\epsilon)h(\beta) + d(\beta) + O(1).$$

Of course, much has been left out of the above discussion. For example, the divisor $g(f) = 0$ is singular for $d \geq 3$ (where f has triple roots or more than one pair of double roots). Also, it is far from clear that the exceptional subvarieties, which hold the key to the truth of the conjecture, all must be linear. Hence, it seems that looking at the coefficients of the minimal polynomial would not be a feasible way of looking at this conjecture, even in its simplest case.

On the positive side, however, the above discussion does show that the conjecture seems to be consistent with itself in a nontrivial way. Also, it indicates how one might reduce part (b) of the conjecture to part (a) with a two-fold covering.

§4. Other Remarks on the General Conjecture

PROPOSITION 5.4.1. *If part (b) of the General Conjecture holds in the special case when $D = 0$, then it holds generally. Likewise, if V is a curve, then the same applies to part (a).*

PROOF: Let V, D, and K be as in Conjecture 5.2.6. Choose rational functions $f_1, \ldots, f_m$ on V such that the intersection of the supports of the principal divisors (f_i) is equal to D. Let n be a positive integer prime to all multiplicities occurring in (f_i), and let V_n be a complete nonsingular variety whose function field is $k(V)(\sqrt[n]{f_1}, \ldots, \sqrt[n]{f_m})$; we may assume that the map $\pi_n \colon V_n \to V$, corresponding to the inclusion of function fields, is a morphism. Let R_n be the ramification divisor of π_n; K_n, the canonical divisor of V_n. Then, for all points $P_n \in V_n$ lying above $P \in V$, we have,

$$h_{K_n}(P_n) \leq d(P_n) + \tfrac{\epsilon}{2} h_A(P) + O(1)$$
$$\leq d(P) + N(R_n, P_n) + \tfrac{\epsilon}{2} h_A(P) + O(1),$$

by the assumption that the General Conjecture holds and by Theorem 5.1.6, respectively. Therefore, by Lemma 5.2.1,

$$(5.4.1.1) \qquad m(R_n, P_n) + h_{\pi_n^* K}(P_n) \leq d(P) + \tfrac{\epsilon}{2} h_A(P) + O(1).$$

But $nR_n \geq (n-1)\pi_n^* D$. Indeed, to see this, it is sufficient to show that, in the notation of Lemma 5.2.2,

$$\sum_{j=1}^{s} a_{ij} \geq n.$$

This follows by the choice of V_n: each of the functions f_i is an nth power on V_n; therefore all components of $\pi_n^* D$ have multiplicity at least n.

Applying this result gives,

$$\left(\frac{n-1}{n}\right) m(D, P) + h_K(P) \leq d(P) + \tfrac{\epsilon}{2} h_A(P) + O(1).$$

Taking n sufficiently large, $\frac{1}{n} m(D, P)$ is less than $\frac{\epsilon}{2} h_A(P)$. The result then follows. $\qquad\qquad\qquad\qquad\qquad\qquad\qquad\qquad\qquad\qquad\qquad\qquad\qquad\quad$ $\square$

REMARK 5.4.2. We note that the conjecture also preserves products. Indeed, let D and E be normal crossings divisors on varieties V and W, respectively. Then $\pi_1^* D + \pi_2^* E$ is a normal crossings divisor on $V \times W$, and $K_{V \times W} = \pi_1^* K_V + \pi_2^* K_W$. Then the General Conjecture gives, for points $P \in V$ and $Q \in W$,

$$m(D, P) + h_{K_V}(P) \leq (\dim V) d(P) + \epsilon h_A(P);$$

$$m(E, Q) + h_{K_W}(Q) \leq (\dim W) d(Q) + \epsilon h_A(Q).$$

These two equations add, termwise, to give the corresponding statement for $V \times W$.

For part (a), the $\dim V$ and $\dim W$ factors are absent. This case is much more difficult, since the appropriate inequality of discriminant terms points in the wrong direction,

$$d((P, Q)) \leq d(P) + d(Q).$$

Equality holds if and only if the fields $k(P)$ and $k(Q)$ are linearly disjoint over $\mathbf{Q}$. But this does not necessarily happen, since, for example, if $V = W$ and $D = E$, then points on the diagonal would have $k(P) = k(Q)$ since $P = Q$. However, the diagonal is a proper Zariski-closed subset, so this merely gives an example in which the Zariski-closed subset must be enlarged to accommodate the extra term in Conjecture 5.2.6. More generally, this situation arises from maps $f: V \twoheadrightarrow W$ with $D \subseteq f^* E$. If $K_w + E$ is almost ample, however, then there are only finitely many such maps, by the function field analogue of the Main Conjecture of Chapter 3. If $K_W + E$ is not almost ample, then Conjecture 5.2.6 (for number fields) is almost vacuous, so again part (a) is additive.

REMARK 5.4.3. We consider how the ramified conjecture applies to points lying above rational points on the ramified cover $\pi \colon \mathbf{P}_1^1 \to \mathbf{P}_2^1$ given by $(x, y) \mapsto (x^2, y^2)$. (The subscripts are used to distinguish between which $\mathbf{P}^1$ we are referring to.) On $\mathbf{P}_2^1$, let $\alpha_1, \ldots, \alpha_n$ be distinct points not equal to zero or infinity. Then Roth's theorem gives,

$$(5.4.3.1) \qquad \sum m(\alpha_i, P) \le (2 + \epsilon) h(P).$$

For $Q \in \mathbf{P}_1^1$, the General Conjecture gives,

$$\sum_i m(\pi^* \alpha_i, Q) \le (1 + \epsilon) h(Q) + \tfrac{1}{2} \log D_{k(Q)},$$

$$(5.4.3.2) \qquad \qquad \le (2 + \epsilon) h(Q)$$

since $d(Q) \le N(R, Q) \le 2h(Q)$. This indicates that the inequality on $d(Q)$ should be very close to equality for points Q on $\mathbf{P}_2^1$ for which $\pi(Q)$ is close to equality in (5.4.3.1). Thus, for example, the largest square factor of x or y in a solution (x, y) of Pell's equation, should be $O(|x|^\epsilon)$. This assertion is supported also by complex analytic considerations, namely the idea that a derivative reduces the order of a zero by one. See Section 6.7 for details.

REMARK 5.4.4. Finally, we note that the special case (5.4.3.2) of the General Conjecture follows from the Main Conjecture by considering bounds on

$$\|w^2 y - x^2 \alpha\|$$

for fixed $\alpha \in k$ and varying w, x, $y \in \mathbf{Z}$. This is done by applying the conjecture to a blown-up projective space in much the same way as was done for Hall's conjecture in Section 4.4. The details are likewise very technical and have been omitted.

§5. Examples: Fermat, abc, Hall, and Hall-Lang-Stark

In this section we will see how the General Conjecture implies several conjectures in number theory that have already been posed. First, it implies that the Fermat conjecture holds for almost all exponents n. Next we consider the Masser-Oesterlé "abc" conjecture, which also implies several other older conjectures. Next follows a shortened derivation of the Hall conjecture over number fields, as well as a generalization due to Lang and Stark, relating to integral points on elliptic curves.

The first two examples use the simplest possible version of the General Conjecture, namely the assertion that

$$h_K(P) \le d(P) + \epsilon\, h_A(P) + O(1)$$

for all algebraic points P on a curve X with canonical divisor K. This is only nontrivial if the genus is at least two; in that case, we can take $A = K$ and replace $1 - \epsilon$ with $1/(1+\epsilon)$, giving

$$(5.5.0.1) \qquad\qquad h_K(P) \le (1+\epsilon)d(P) + O(1)$$

The third example uses the proximity term $m(D, P)$ as well, and the last example uses this machinery on a surface instead of a curve. Thus we use successively more complicated aspects of the conjecture in each example, running into more complications in each case.

EXAMPLE 5.5.1 (ASYMPTOTIC FERMAT). We first see how the General Conjecture implies that Fermat's Last Theorem holds for almost all exponents n. Let X be the curve $x^4 + y^4 = z^4$, and let a, b, c be positive integers for which $a^n + b^n = c^n$. Then $P = [a^{n/4} : b^{n/4} : c^{n/4}]$ is a point on X with $h_K(P) = (n/4)\log\max(a,b,c)$ but $d(P) \le \log 2abc$. Inequality (5.5.0.1) would then imply a bound on n for which there is a rational solution $[a : b : c]$. It would even suffice to prove $h_K(P) \le Cd(P) + O(1)$ for any fixed C, so that by Parshin's trick it would suffice to know (5.7.7) below, with any constant in place of $g/2$.

EXAMPLE 5.5.2 (THE "ABC" CONJECTURE). "Fermat for almost all n" is also a consequence of the following "abc" conjecture.

CONJECTURE 5.5.2.1 (MASSER, OESTERLÉ). *If $a, b, c \in \mathbf{Z}$ are relatively prime integers satisfying $a + b = c$, and if $N = Conductor(abc)$ is the product of all primes dividing abc, then for all $\epsilon > 0$,*

$$\max(|a|, |b|, |c|) \ll N^{1+\epsilon}$$

where the constant in $\ll$ depends only on ϵ.

This conjecture is also a consequence of Conjecture 5.2.6. Indeed, choose n large enough so that $\epsilon' = (n-3)\epsilon - 2$ is positive and let X_n be the projective curve $X^n + Y^n = Z^n$. Each triple (a, b, c) gives a point $[x : y : z] = [\sqrt[n]{a} : \sqrt[n]{b} : \sqrt[n]{c}]$ on X_n of height $\frac{1}{n}\log\max(|a|, |b|, |c|)$. By the adjunction formula, the height relative to the canonical divisor on X_n is then $\frac{n-3}{n} \log \max(|a|, |b|, |c|)$. The normalized logarithmic discriminant is $d(P) \le \frac{n-1}{n} \log N + O(1)$. Applying (5.5.0.1) to X_n then gives,

$$\frac{n-3}{n} \log \max(|a|, |b|, |c|) < \left(\frac{n-1}{n} + \frac{\epsilon'}{n} \right) \log N + O(1).$$

The abc conjecture follows immediately from this.

This "abc" conjecture also implies Hall's conjecture over $\mathbf{Z}$; in the next example we derive the version over number fields, 4.4.2, with much less work than in Section 4.4. For more details on the abc conjecture, see the Appendix to this chapter.

EXAMPLE 5.5.3 (HALL'S CONJECTURE). For this example we will need the $m(D, P)$ term in Conjecture 5.2.6, as well as the terms present in (5.5.0.1). (We have already seen, though, that (5.5.0.1) actually implies the corresponding statement with the $m(D, P)$ term—see Proposition 5.4.1). In this example, as in Section 5.3, the curve in question is $\mathbf{P}^1$. The general idea is that, since $\mathbf{G}_m$ acts on the variety V constructed in Section 4.4, one would like to reduce Hall's conjecture to a diophantine statement on a curve representing the $\mathbf{G}_m$-orbits of V. The problem is that $\mathbf{G}_m$ does not act freely on V. Indeed, since the action is by $u \cdot (x, y) = (u^2 x, u^3 y)$, ± 1 acts trivially on $(x, 0)$ and cube roots of unity act trivially on $(0, y)$. Thus a space of orbits of V does not exist as an algebraic curve, but it will exist for a suitably ramified cover of V.

Instead of explicitly constructing this cover of V, we will construct the orbit space directly. This space is $\mathbf{P}^1$; a point $(x, y) \in \mathbf{A}^2$ corresponds to the point $(\sqrt{x}, \sqrt[3]{y}) = (a, b)$ on $\mathbf{P}^1$. If $y^2 - x^3$ is small, then $\sqrt{x}/\sqrt[3]{y}$ will be close to a 6th root of unity. Thus, for all $v \in S$,

$$\max(1, \|y^2 - x^3\|_v) = \max(1, \|b^6 - a^6\|_v)$$

$$\gg \left(\frac{\max(\|a\|_v, \|b\|_v, 1)}{\max(\|a\|_v, \|b\|_v)} \right)^6 \prod_{i=0}^{5} \|a - b\varsigma^i\|_v,$$

where ς is a primitive 6th root of unity. Hence we apply the General Conjecture to the element $a/b \in \mathbf{P}^1$, relative to the points ς^i. This gives,

$$(5.5.3.1) \qquad \sum_{i=0}^{5} m\left(\varsigma^i, \frac{a}{b}\right) < (2 + \epsilon) h(a, b) + d(a, b).$$

Now $Q(\sqrt{x})/Q$ has a discriminant of at most the multiplicative height $4H(x)$ for $x \in Q$. This holds over any number field k if one uses the absolute height: $d(k(\sqrt{x})) - d(k) \le h(\sqrt{x}) + \log 2$. Similarly, in the case of cubic fields, the discriminant of $Q(\sqrt[3]{y})/Q$ is at most $27H(y)^2$, hence $d(k(\sqrt[3]{y})) - d(k) \le 2h(\sqrt[3]{y}) + \log 27$. Therefore, if x and y are relatively prime integers,

$$d((a,b)) \le h(a) + 2h(b) + O(1)$$
$$\le 3h(a/b) + O(1).$$

With (5.5.3.1) this gives,

$$\sum_{i=0}^{5} m(\varsigma^i, \frac{a}{b}) \le (5+\epsilon)h(a/b);$$

In the following, let w denote places of $k(a,b)$. By abuse of notation, we write $w \mid S$, meaning that w lies over some $v \in S$. The above inequality then reads,

$$\sum_{w|S} \sum_{i=0}^{5} -\log\min\left(1, \|\tfrac{a}{b} - \varsigma^i\|_w\right) \le (5+\epsilon) \sum_{w} \log\max(\|a\|_w, \|b\|_w)$$

$$\sum_{w|S} \sum_{i=0}^{5} -\log \frac{\|a - b\varsigma^i\|_w}{\max(\|a\|_w, \|b\|_w)} \le (5+\epsilon) \sum_{w} \log\max(\|a\|_w, \|b\|_w)$$

Since x and y are S-integers, so are a and b; therefore,

$$h(a/b) = \frac{1}{[k(a,b):Q]} \sum_{w|S} \log\max(\|a\|_w, \|b\|_w).$$

Thus the above inequality becomes,

$$\sum_{w|S} \sum_{i=0}^{5} -\log \frac{\|a - b\varsigma^i\|_w \max(\|a\|_w, \|b\|_w, 1)}{\max(\|a\|_w, \|b\|_w)}$$

$$\le (5+\epsilon) \sum_{w|S} \log\max(\|a\|_w, \|b\|_w)$$

$$- 6 \sum_{w|S} \log\max(\|a\|_w, \|a\|_w, 1);$$

$$\sum_{w|S} \log \max(1, \|y^2 - x^3\|_w) > (\tfrac{1}{6} - \epsilon) \sum_{w|S} \log \max(\|x^3\|_w, \|y^2\|_w, 1)$$

Since x and y are integers, this becomes the usual Hall conjecture,

$$h(y^2 - x^3) > (\tfrac{1}{6} - \epsilon)h([x^3 : y^2 : 1]).$$

For fixed positive integers m, n the same derivation yields,

$$h(x^m - y^n) > (1 - \tfrac{1}{m} - \tfrac{1}{n} - \epsilon)h([x^m : y^n : 1]),$$

which also has been conjectured to hold ([**L 6**, p. 213]) and proved over function fields ([**Sil 1**], [**Mas 1**], [**Mas 2**]).

REMARK 5.5.4. Our use of the General Conjecture is very similar to Silverman's use of the Hurwitz genus formula in proving Hall's conjecture in the function field case [**Sil 1**]. Other applications of the genus formula can similarly be translated.

EXAMPLE 5.5.5 (THE HALL-LANG-STARK CONJECTURE). This generalization of Hall's conjecture concerns elliptic curves.

CONJECTURE 5.5.5.1 (LANG, STARK; [**L 10**]). *If a, b are integers, if the equation,*

(5.5.5.2) $$y^2 = x^3 + ax + b$$

describes an elliptic curve (i. e. is nonsingular), and if $(x, y) \in \mathbf{Z}^2$ is a point on this curve, then

(5.5.5.3) $$|x| \ll \max(|a|^3, |b|^2)^{5/3+\epsilon}.$$

The constant in $\ll$ should depend only on ϵ.

Lang originally posed the conjecture with an unknown exponent; then Stark suggested, on probabilistic grounds, that the exponent should be $5/3$.

This example will apply the General Conjecture to a surface instead of a curve. It is therefore no surprise that exceptional subvarieties (at least two curves in this case) will play a role. We can therefore conclude only that Conjecture 5.2.6 implies 5.5.5.1 modulo a finite number of exceptional curves.

For each $a, b, x, y \in \mathbf{Z}$ satisfying (5.5.5.2), let a corresponding point $P \in \mathbf{P}^2$ be defined in homogeneous coordinates as,

$$P = [u : v : w]$$
$$= [\sqrt[3]{y} : \sqrt{x} : \sqrt[4]{a}].$$

Then the equation $b = 0$ becomes the curve,

$$D: u^6 = v^6 + v^2 w^4.$$

We will want to apply the conjecture to D, but first it is necessary to check that the singularities are not too bad. In fact, D is singular at the point $[0 : 0 : 1]$. At that point, two smooth branches of D meet with intersection number three. Blowing up twice reduces this intersection number to one:

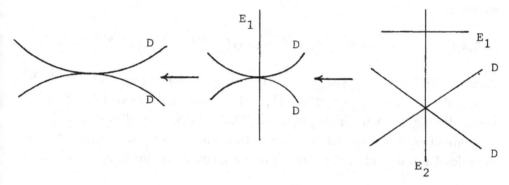

Let $\pi: V \to \mathbf{P}^2$ denote this blowing-up. Let $D^{\#}$ be the strict transform of D (i. e. the prime divisor on V whose image is D), and let W denote the divisor $w = 0$. We will apply the General Conjecture to the divisor $D^{\#} + E_1^{\#} + W$, which is a normal crossings divisor on V. Part (b) applies in this case because the points P lie over $\mathbf{P}^2(\mathbf{Q})$ under the map $[u : v : w] \mapsto [u^6 : v^6 : v^2 w^4]$. Thus,

$$m(D^{\#} + E_1^{\#} + W) + h_{K_V}(P) < \epsilon h_{\pi^* \mathcal{O}(1)}(P) + d(P) + O(1),$$

outside of a finite union of curves. But also,

$$K_V = \pi^* \mathcal{O}(-3) + E_1^{\#} + 2E_2;$$
$$\pi^* D = D^{\#} + 2E_1^{\#} + 4E_2;$$
$$m(E_i, P) \leq h_{E_i}(P) + O(1), \qquad i = 1, 2;$$

and

$$d(P) \leq \log |y|^{2/3} |x|^{1/2} |a|^{3/4} + O(1).$$

Combining these inequalities gives,

$$m(\pi^* D + W - 2E_2, P) < (3 + \epsilon) h_{\pi^* O(1)}(P) + \log |y|^{2/3} |x|^{1/2} |a|^{3/4} + O(1).$$

Using $\pi^* E_1 \geq E_2$ and substituting for definitions, this becomes,

$$- \log \frac{|w||u^6 - v^6 - v^2 w^4|}{\max(|u|, |v|, |w|)^7} + 2 \log \frac{\max(|u|, |v|)}{\max(|u|, |v|, |w|)}$$
$$< (3 + \epsilon) \log \max(|u|, |v|, |w|) + \log |y|^{2/3} |x|^{1/2} |a|^{3/4} + O(1).$$

Since we are interested in points where $|x| > |a|^5$ (i. e. $|v| > |w|^{10}$), the second term on the left can be ignored. Translating into the original variables, we have,

$$(5.5.5.4) \qquad |a||b||x|^{1/2}|y|^{2/3} \gg \max(|y|^{1/3}, |x|^{1/2}, |a|^{1/4})^{4-\epsilon}.$$

Now, if (5.5.5.3) is false, then (ignoring ϵ for now), $|a| \ll |x|^{1/5}$; $|b| \ll |x|^{3/10}$. Hence, $|y|^2 \lll |x|^3$, and (5.5.5.4) reduces to $|a||b| \gg |x|^{1/2}$, contradicting what we have just said. Thus (5.5.5.4) implies (5.5.5.3).

Remember that the General Conjecture only asserts that (5.5.5.4) holds outside of a finite union of curves. Two such curves are given by the equation,

$$u^6 = v^6 + v^2 w^4 + c w^6,$$

where $c = \pm 2i/3\sqrt{3}$. These curves correspond to the following solution of (5.5.5.2):

$$(t^3 + \tfrac{3}{2}t)^2 = (t^2 + 1)^3 - \tfrac{3}{4}(t^2 + 1) - \tfrac{1}{4}.$$

But the corresponding cubic curve is singular and therefore rational, not elliptic. It may be possible that there are other exceptional curves, but I do not know of any.

It is possible to derive an inequality similar to (5.5.5.4) by using the Main Conjecture of Chapter 3, but it is much harder, just as it was with the original Hall conjecture. Indeed, it is necessary to work on three-folds, and perform eight blowings-up, each of which is twice as complicated as blowing up on a surface.

§6. The (split) Function Field Case

In this section, we prove Conjecture 5.2.6 in the special case of isotrivial curves over one-dimensional function fields, relative to divisors which are also isotrivial. In this case, we can derive effective explicit estimates for the bounds involved. This result is also valid in characteristic $p > 0$. "Isotrivial" means that, after a finite base change, the curve and divisor are defined over the constant field. Since, in our case, the quantities m, N, d, etc. change by only a bounded amount under base change, the question immediately reduces to the split case.

Throughout this section, let k be an algebraically closed field (the constant field), and let B be a complete nonsingular curve defined over k, so that $k(B)$ is the function field. Let S be a finite set of closed points of B. Let C be a complete nonsingular curve over $k(B)$, and let X be a nonsingular complete surface with a map $\pi\colon X \to B$ whose generic fibre is C.

In order to discuss the General Conjecture in the function field case, it is necessary to define the Nevanlinna machinery ($m(D,P)$, etc.) for function fields. This is fairly easy, since Weil functions over non-archimedean places have already been defined in terms of intersection pairings in the proof of Proposition 1.4.7. Because there are minor differences, however, we repeat the definitions here.

Let P be a point on C, and let $s\colon B \to X$ be the corresponding section of π. We can also think of P as the image of s. Let D_0 be a divisor on C, and let D be its closure on X. Then we can define Weil functions,

(5.6.1)
$$\lambda_{D,v}(P) = (P.D)_v$$
$$= \text{length}_{\mathcal{O}_{B,v}}(\mathcal{O}_{B,v}/g(s(v)))$$

for a closed point $v \in B$, if D is locally defined by the function g. Then

$$m(D,P) = \sum_{v \in S}(P.D)_v;$$

$$N(D,P) = \sum_{v \notin S}(P.D)_v;$$

$$h_D(P) = \sum_{v}(P.D)_v = (P.D).$$

If P is an algebraic point defined, say, over B', then we take intersections on $X' = X \times_B B'$, by pulling back D to a divisor D' on X' and letting

$$\lambda_{D,w}(P) = \frac{(P.D')_w}{[k(B') : k(B)]}$$

for some point $w \in B'$ lying over $v \in B$.

Theorem 5.1.6 also works, provided $d(k)$ (or, in this case, $d(B')$), is defined correctly. In place of (5.1.5), we have,

$$d(B') - d(B) = \frac{1}{[k(B') : k(B)]} \sum_{w \in k(B')} \text{length}_w \, \Omega_{B'/B}$$

$$= \frac{\deg R_{B'/B}}{[k(B') : k(B)]}$$

so that it is possible to let,

$$d(B') = \frac{2g(B') - 2}{[k(B') : k(B)]}.$$

In the finite characteristic case, recall that the maps $V \xrightarrow{\pi} W$ and $B' \to B$ are required to be separable.

All of the above discussion is valid generally; i. e., for nonsingular complete varieties over one-dimensional function fields of any characteristic. We now restrict to the special case where all curves and divisors are defined over k. In that case, we can take $X = B \times C$, and Lemma 5.1.8 applies to all places outside of S. Also, there is a canonical choice of Weil function, so that by Remark 5.1.4, it makes sense to refine the inequality of Theorem 5.1.6 to,

$$(5.6.2) \qquad d(Q) - d(P) \le N(R, Q) + |S|(1 + \text{ord}_p n)$$

where p is the characteristic of k (if nonzero) and n is the degree of the map $\pi : V \to W$. Likewise, the General Conjecture for curves can be proved in the split case, giving concise explicit bounds.

THEOREM 5.6.3. *Let C be a complete nonsingular curve over $k(B)$ which is defined over k; and let D be an effective divisor on C which is also defined over k and has no multiple points. Let P be an algebraic point on C; assume that (i) P is not defined over the constant field; and (ii) P is defined over a separable extension of $k(B)$. Then*

$$m(D, P) + h_K(P) \le d(P) + |S|.$$

REMARKS.

(a). Restriction (ii), above, is almost as reasonable as (i). Indeed, a point defined over an inseparable field is everywhere tangent to a constant.

(b). This generalizes somewhat a result of Mason [**Mas 1**], that all non-constant solutions of the unit equation

$$u + v = 1; \qquad u, v \in \mathcal{O}(B \setminus S)^*,$$

have $h(u) \leq 2g - 2 + |S|$. This follows from (5.6.3) by letting $C = \mathbf{P}^1$ and $D = [0] + [1] + [\infty]$. Silverman [**Sil 2**] has an independent proof, of which our proof is an extension. He also has an example in which the bound is best possible: let $C = \mathbf{P}^1$ and $n > 0$. Let ς be a primitive nth root of unity. Then

$$X^n + (1 - X^n) = 1$$

is a solution of the unit equation of height n, where $g = 0$ and $S = \{0, \infty, 1, \varsigma, \ldots, \varsigma^{n-1}\}$ has $n + 2$ elements.

Osgood [**Os 2**] also has an effective Roth theorem for function fields which is valid without the above restrictions on D. His bound, although better than Roth's, is not as sharp as the above. Furthermore, it is limited to one absolute value.

PROOF: **Case I.** $D = 0$, $S = \emptyset$. Let g be the genus of C. Then what we want to show is that,

$$\frac{(2g - 2) \deg f}{\deg \phi} \leq \frac{2g' - 2}{\deg \phi}$$

where $f \colon B' \to C$ determines the point on C and $\phi \colon B' \to B$ corresponds to the inclusion $k(B) \subseteq k(B')$. But this is an easy consequence of the Hurwitz genus formula.

Case II. D, S arbitrary. This is a simpler version of the proof of Proposition 5.4.1. Fix a geometric point $P \in C$. Let n, $\pi_n \colon C_n \to C$, R_n, K_n, and P_n be as in the proof of 5.4.1 (substituting C_n and C for V_n and V). Assume also that n is relatively prime to the characteristic of k, if the latter is nonzero. The sharper inequalities from Case I and from (5.6.2) give the sharper inequality,

$$m(R_n, P_n) + h_{\pi_n^* K}(P_n) \leq d(P) + |S|,$$

in place of (5.4.1.1). Again, $nR_n \geq (n-1)\pi_n^* D$; therefore, by functoriality,

$$\left(\frac{n-1}{n}\right) m(D, P) + h_K(P) \leq d(P) + |S|.$$

Letting $n \to \infty$, (5.6.3) follows. $\qquad \qquad \square$

§7. The ramification term in the Conjecture using (1,1) forms

In this section we add a ramification term to Theorem 4.3.4, giving a discriminant term in Conjecture 4.3.5. Applying this conjecture to the moduli space of abelian varieties then gives a conjecture which generalizes the Shafarevich conjecture.

DEFINITION 5.7.1. *Let $f: C \to V$ be a holomorphic function to a complete nonsingular variety V.*

(a). *Let ω be a $(1,1)$ form on V. Then the <u>characteristic function of f relative to ω</u> is,*

$$T_\omega(r) = \int_0^r \left(\int_{B[t]} f^*\omega \right) \frac{dt}{t}.$$

If $\omega = c_1(\mathcal{L})$ for some metrized line bundle $\mathcal{L}$ (cf. Section 4.3), then the above definition corresponds to the usual definition of $T_\mathcal{L}$.

(b). *Let $n_1(r)$ be the number of ramification points of f (counted with multiplicities) in the disc $|z| < r$. Assume $n_1(0) = 0$. Then let*

$$N_1(r) = \int_0^r n_1(r) \frac{dr}{r}.$$

THEOREM 5.7.2. *Let V be a variety as above and let D be a normal crossings divisor on V. Let ω be a holomorphic $(1,1)$ form on $V \backslash D$ whose holomorphic sectional curvatures are bounded from above by $-c < 0$; i. e. for any holomorphic map $f: U \to V$ ($U \subseteq C$ is an open subset), we have,*

$$\text{Ric } f^*\omega \geq c f^*\omega.$$

Let $\epsilon > 0$. Then for any holomorphic map $f: C \to V \backslash D$ unramified at the origin,

$$T_\omega(r) + \tfrac{1}{c} N_1(r) \leq \epsilon T_\omega(r). \qquad \qquad //$$

PROOF: We recall the terminology of currents as in [G] or [G-K]. On an n-dimensional complex manifold, a (p,q) current is a linear functional on the space of compactly supported C^∞ $(n-p, n-q)$ forms on V. Currents arise in Nevanlinna theory in one of the following ways:

(i). A C^∞ (p,q) form ψ defines a current $[\psi]$ by the rule,

$$[\psi](\phi) = \int_V \psi \wedge \phi.$$

(ii). A divisor D defines a $(1,1)$ current $[D]$ by,

$$[D](\phi) = \int_D \phi.$$

(iii). Finally, a (p, q) current T defines a $(p+1, q)$ current ∂T by,

$$\partial T(\phi) = (-1)^{p+q} T(\partial \phi).$$

Similarly, we define $\overline{\partial} T$, dT, and $d^c T$. The sign is arranged so that $d[\phi] = [d\phi]$ by Stokes' theorem. One of Griffiths' major contributions to Nevanlinna theory has been the introduction of currents, in particular noting that the theorems come from examining $dd^c[\phi] - [dd^c\phi]$ for singular forms ϕ.

Now to prove (5.7.2), we let ς be the function on $\mathbb{C}$ for which,

$$f^*\omega = \varsigma \, dz \wedge d\bar{z}.$$

Also let R be the ramification divisor of f. As in [G-K, 9.26], we have the equation of $(1,1)$ currents,

$$
\begin{aligned}
dd^c[\log \varsigma] &= [R] + [\operatorname{Ric} f^*\omega] \\
&\geq R + c[f^*\omega],
\end{aligned}
$$

(5.7.3)

by the curvature assumption. Let

$$\mu(r) = \int_{\partial B[r]} \log \varsigma \sigma$$

where $\sigma = d\theta/2\pi$ in polar coordinates. Then doubly integrating (5.7.3) gives, as in [G-K, 6.23],

(5.7.4)
$$\mu(r) \geq N_1(r) + cT_\omega(r).$$

But by concavity of the logarithm, [G-K] 7.22, and 7.26, respectively,

$$
\begin{aligned}
\mu(r) &\leq \log \int_{\partial B[r]} \varsigma \sigma \\
&= \log \frac{d^2 T(r)}{ds^2} \qquad \left(\frac{d}{ds} = r\frac{d}{dr} \right) \\
&\leq \epsilon T(r) + O(\log r).
\end{aligned}
$$

With (5.7.4) this gives the desired estimate. □

This theorem omits the assumption from (4.3.4) that ω is large (4.3.3). This is only needed for the conclusion that f must be rational. Indeed, from Theorem 5.7.2 we have,

$$T_\omega(r) \leq O(\log r).$$

If $\mathcal{L}$ is a line bundle satisfying the conditions of Definition 4.3.3, then $T_{\mathcal{L}}(r) \ll T_\omega(r)$; hence f must be rational.

In the above discussion, the inclusion of $N_1(r)$ seems rather silly since it is always positive. However, in the number field case, $N_1(r)$ becomes $-d(P)$, so the term takes on significance here.

CONJECTURE 5.7.5. *Let ω, V, D, c, and ϵ be as above, and let A be an almost ample divisor on V. Also assume that $\omega \geq c_1(\mathcal{L})$ for some metrized line bundle $\mathcal{L}$ on V. Then for all D-integralizable sets of closed points P of V of bounded degree over $\mathbf{Q}$,*

$$h_{\mathcal{L}}(P) \leq \tfrac{1}{c}d(P) + \epsilon h_A(P) + O(1).$$

Note that, as in the Conjecture using $(1,1)$ forms, there is no exceptional Zariski-closed subset.

EXAMPLE 5.7.6. Let $\mathcal{A}_{g,n}$ be the moduli space of principally polarized abelian varieties of dimension g with level-n structure. By the toroidal compactification, we have $\mathcal{A}_{g,n} = \overline{\mathcal{A}}_{g,n} \setminus D$ where D is a normal crossings divisor. Let K denote the canonical divisor of $\overline{\mathcal{A}}_{g,n}$. Faltings [**F**] has defined a height h_F of a point on $\mathcal{A}_{g,n}$. It is equivalent (up to a small error term) to $h_{\mathcal{L}}$ where W is a vector bundle of rank g and

$$\mathcal{L} = \wedge^g W.$$

We also have,

$$\Omega^1[\log D] = S^2(W).$$

By 4.1.1 and linear algebra,

$$K_V + D = \wedge^{\max}\Omega^1[\log D]$$
$$= (g+1)\mathcal{L}.$$

Thus $h_F \sim h_{K+D}/(g+1)$.

For n sufficiently large, $\mathcal{A}_{g,n}$ is a scheme over $\operatorname{Spec}\mathbf{Z}$; by base change we will view it as a scheme over $\operatorname{Spec}\mathcal{O}_{k,S}$. Then, in the language of schemes, an integral point is a section of the map to $\operatorname{Spec}\mathcal{O}_{k,S}$, which corresponds to a principally polarized semiabelian variety with level-n structure and good reduction outside S. An algebraic integral point corresponds to such an abelian variety defined over a finite extension of $\mathcal{O}_{k,S}$ with good reduction at places above $M_k - S$. By Proposition 1.4.7, such abelian varieties correspond to D-integral points on $\overline{\mathcal{A}}_{g,n}$.

Let ψ be the invariant volume form on $\mathcal{A}_{g,n}$. We wish to apply Conjecture 5.7.5 to $\omega = \operatorname{Ric}\psi$. ψ comes from a singular metric on $K + D$; by the remark following 5.7.1, we would have $T_\omega = T_{K+D}$ if ψ corresponded to a smooth metric. Although the metric is not smooth in this case, the singularities are bounded by a power of the logarithm of the distance to D; by [C-G, 5.15] (or [F, Lemma 3]), such singularities do not make a substantial difference. As for the constant c in (5.7.5), it is not hard to find tangent vectors whose sectional curvature is $-2/(g+1)$. By [K, p. 41], the sectional curvatures cannot vary by more than a factor of g; hence we may take $c = 2/g(g+1)$. This gives,

$$h_{K+D}(P) \le \frac{g(g+1)+\epsilon}{2}d(P)$$

Therefore, for semiabelian varieties A defined over fields $k(A)$ of bounded degree and having good reduction outside S, one would have,

(5.7.7) $$h_F(A) \le (\tfrac{g}{2} + \epsilon)d(k(A)) + O(1).$$

The General Conjecture gives roughly the same answer. Indeed, for semiabelian varieties A with good reduction outside S, $m(D, A) = h_D(A)$, so

$$m(D, A) + h_K(A) = h_{K+D}(A) + O(1).$$

Then 5.2.6(b) conjecturally gives,

$$(g + 1)h_F(A) \le (\dim \mathcal{A}_{g,n} + \epsilon)d(A) + O(1)$$

outside of some Zariski-closed subset of $\mathcal{A}_{g,n}$. This agrees with 5.7.7 since $\dim \mathcal{A}_{g,n} = g(g+1)/2$.

Arakelov [**Ar 1**] has obtained this bound in the function field case for jacobian abelian varieties. Indeed, if C is a curve of genus g over a base B

of genus q with at most s points of bad reduction, and g_0 is the degree of the constant part of the jacobian, then

$$\deg \det \Omega^1_{C/B} \le (q - 1 + \tfrac{s}{2})(g - g_0).$$

Raynaud [**Ray**] has proved a similar bound over number fields, for semi-abelian varieties within an isogeny class. The constant is larger, however, and also depends on $[k(A) : \mathbf{Q}]$.

Appendix ABC. The Masser-Oesterlé "abc" conjecture.

In this appendix, we collect, for the convenience of the reader, some implications concerning the "abc" conjecture which have not appeared elsewhere. These implications are the following:

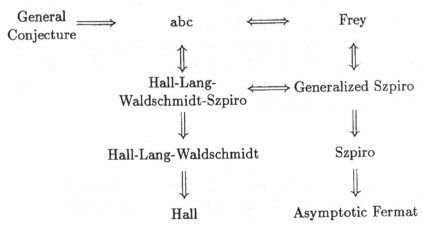

Some of the above conjectures have not been introduced yet; they will be stated below. The "abc" conjecture itself has already been stated in (5.5.2), but we repeat the statement in full generality:

CONJECTURE. *Let k be a global field and S a finite set of places of k containing the archimedean places. Let $N_v = 1$ in the function field case and $N_v = \frac{f \log p}{[k:\mathbf{Q}]}$ in the number field case, if the residue field of v is $\mathbf{F}_{p^f}$. Then for all $\epsilon > 0$ and all $a, b, c \in \mathcal{O}_k$ such that $a + b = c$, we have,*

$$h([a : b : c]) < (1 + \epsilon) \sum_{\substack{v \in S \\ v(abc) > 0}} N_v + O(1),$$

where the constant depends only on ϵ, k, and S.

On the left-hand side of the above inequality, the height refers to projective coordinates. The right-hand side could be taken as a definition of the

<u>conductor</u> of the quantity abc, written conductor(abc). The multiplicative equivalent would be capitalized, so that

$$\text{conductor}(x) = \frac{1}{[k:\mathbf{Q}]} \log \text{Conductor}(x).$$

Mason has proved this in the function field case (with $\epsilon = 0$ and $O(1)$ explicitly given) ([**Mas 1**], [**Mas 2**]). This also follows from Example 5.5.2 and Theorem 5.6.3, but we include here a short proof over $\mathbf{C}[t]$: dividing by c, we have $u + v = 1$, with $u, v \in \mathbf{C}(t)$. Taking derivatives gives $u' = -v'$; then

(5.A.0.1)
$$\frac{u}{v} = -\frac{v'}{v} \bigg/ \frac{u'}{u}.$$

But the logarithmic derivative of

$$\prod (z - \alpha)^{m(\alpha)}$$

is

$$\sum \frac{m(\alpha)}{z - \alpha}$$

Counting degrees on the right of (5.A.0.1) then gives the desired bound.

$\square$

We now prove the various implications, restricting to the case over $\mathbf{Z}$ for simplicity.

5.A.1. ABC $\implies$ HALL-LANG-WALDSCHMIDT-SZPIRO.

CONJECTURE (H-L-W-S). *Let A, B be fixed integers and let $\epsilon > 0$. Then for all $m, n \in \mathbf{Z}$ satisfying,*

(5.A.1.1)
$$Ax^m + By^n = z \neq 0,$$

we have,

(5.A.1.2)
$$\text{Conductor}(z)^{1+\epsilon} \gg |xy|^{\frac{mn-(m+n)}{m+n}}.$$

Also,

(5.A.1.3)
$$\text{Conductor}(z)^{1+\epsilon} \gg |z|^{\frac{mn-(m+n)}{mn}}.$$

The constants implicit in $\gg$ should depend only on A, B, and ϵ. Over number fields,

$$\text{Conductor}_S(z)^{1+\epsilon} \gg \prod_{v \notin S} \max\left(\|xy\|^{\frac{mn-(m+n)}{m+n}}, \|z\|^{\frac{mn-(m+n)}{mn}} \right).$$

Indeed, applying the abc conjecture to (5.A.1.1) gives,

$$|x|^m \ll (|xy|\,\text{Conductor}(z))^{1+\epsilon}; \qquad |y|^n \ll (|xy|\,\text{Conductor}(z))^{1+\epsilon};$$

therefore,

$$|x|^{mn} \ll |xy|^{n+n\epsilon}\,\text{Conductor}(z)^{n+n\epsilon};$$

$$|y|^{mn} \ll |xy|^{m+m\epsilon}\,\text{Conductor}(z)^{m+m\epsilon};$$

$$|xy|^{\frac{mn-(m+n)(1+\epsilon)}{(m+n)(1+\epsilon)}} \ll \text{Conductor}(z).$$

Now, if $\frac{1}{m} + \frac{1}{n} \geq 1$, then the exponent in (5.A.1.2) is negative, and there is nothing to prove. Otherwise, it is an exercise in algebra to show that, given $\epsilon' > 0$ there exists $\epsilon > 0$ such that,

$$\frac{mn - (m+n)(1+\epsilon)}{(m+n)(1+\epsilon)} > \frac{mn - (m+n)}{(m+n)(1+\epsilon')}$$

independently of m and n. This shows that (abc) $\Longrightarrow$ (5.A.1.2). The abc conjecture applied to (5.A.1.1) also gives,

$$|xy| \gg \frac{z^{1-\epsilon}}{\text{Conductor}(z)};$$

combining this with (5.A.1.2) gives (5.A.1.3) (after adjusting ϵ). $\square$

5.A.2. HALL-LANG-WALDSCHMIDT-SZPIRO $\Longrightarrow$ GENERALIZED SZPIRO.

CONJECTURE. *For an elliptic curve $E/\mathbf{Q}$ in minimal model, we have,*

$$\max(|\Delta|, |g_2^3|) < c \cdot \text{Conductor}(E)^{6+\epsilon}$$

for some positive constant $c = c_\epsilon$ independent of E.

(Szpiro originally conjectured this with 6 replaced with an undetermined constant and only $|\Delta|$ on the left-hand side.)

Applying the Hall-Lang-Waldschmidt-Szpiro conjecture to $\Delta = g_2^3 - 27g_3^2$ gives,

$$|g_2 g_3|^{1/5} \ll \text{Conductor}(\Delta)^{1+\epsilon};$$

$$|\Delta|^{1/6} \ll \text{Conductor}(\Delta)^{1+\epsilon}$$

But, we must have $|g_2|^3 \gg\ll |g_3|^2$ or $|g_2|^3 \gg\ll |\Delta|$. In the first case, $|\Delta|^{1/6} \ll |g_2|^{1/2} \ll |g_2 g_3|^{1/5}$; in the second case, $|g_2|^{1/2} \ll |\Delta|^{1/6}$. $\square$

5.A.3. GENERALIZED SZPIRO $\Longrightarrow$ FREY.

CONJECTURE (FREY) [**Fr 2**]. *Let k be a number field, and let S be a finite set of places of k containing the archimedean places. Let $\epsilon > 0$. Then, for all elliptic curves E/k such that $v(\Delta_E) \equiv 0 \pmod 6$ whenever $v(j_E) \geq 0$ and $v \notin S$, we have,*

$$h(j_E) \leq (6 + \epsilon)\,\text{conductor}(\text{den}(j_E)) + O(1),$$

where $\text{den}(j_E)$ denotes the denominator of j_E and $O(1)$ depends only on k, S, and ϵ.

The conditions on j_E and Δ are equivalent to requiring that E can be written in Weierstrass form with g_2 and g_3 relatively prime integers. Then $\Delta_E = \text{den}(j_E)$, and it easily follows that this is equivalent to the Generalized Szpiro conjecture. $\qquad\square$

5.A.4. FREY $\Longrightarrow$ ABC.

Let $a + b = c$, and consider the Frey elliptic curve,

$$y^2 = x(x + a)(x - b).$$

Changing variables, this becomes,

$$y^2 = 4x^3 - \tfrac{4}{3}(a^2 + ab + b^2)x - \tfrac{4}{27}(b - a)(2a + b)(2b + a).$$

Except for primes above 2 and 3, g_2 and g_3 are relatively prime integers. Indeed, any common factor would have to divide $\Delta = (abc)^2$; considering separately the cases $p \mid a$, $p \mid b$, $p \mid c$ shows that this is not so.
Now

$$\begin{aligned}
h([a : b : c]) &\leq \tfrac{1}{6}h(j_E) + O(1)\\
&< (1 - \epsilon)\,\text{conductor}(\Delta_E) + O(1)\\
&< (1 - \epsilon)\,\text{conductor}(abc) + O(1). \qquad\square
\end{aligned}$$

See also [**Fr 1**, p. 24].

5.A.5. HALL-LANG-WALDSCHMIDT-SZPIRO
$\Longrightarrow$ HALL-LANG-WALDSCHMIDT
$\Longrightarrow$ HALL.

CONJECTURE (HALL-LANG-WALDSCHMIDT). *Fix $\epsilon > 0$. Then for all $m, n, x, y \in \mathbf{Z}$ with $x^m \neq y^n$,*

$$|x^m - y^n| \gg \max(|x|^m, |y|^n)^{1 - \frac{1}{m} - \frac{1}{n} - \epsilon}$$

with the constant depending only on ϵ.

We may assume that $|x|^m \gg\ll |y|^n$; then this conjecture follows from (5.A.1.2). The Hall conjecture is a special case of this conjecture.

5.A.6. ABC $\Longrightarrow$ ASYMPTOTIC FERMAT.

If $u^p + v^p = w^p$, then

$$\begin{aligned}
\log \max(|u|, |v|, |w|) &= \tfrac{1}{p} \log \max(|u^p|, |v^p|, |w^p|) \\
&< \tfrac{1}{p} \operatorname{conductor}(u^p v^p w^p)^{1+\epsilon} + O(1) \\
&< \tfrac{3}{p} \log \max(|u|, |v|, |w|) + O(1). \qquad \square
\end{aligned}$$

The original Szpiro conjecture also applies Asymptotic Fermat, by combining the above methods with those of Section 5.A.4.

Chapter 6
Approximation to Hyperplanes

In this chapter, instead of equidimensional holomorphic maps, we will consider holomorphic maps $C \to V$. In this case, defect relations are much harder to prove—in fact, they are only known when V is a curve or when $V = P^n$ and the divisor D is a union of hyperplanes in general position. Either by coincidence or by fate, these two cases are also the only defect relations proved for number fields. In this chapter we will prove the following theorem.

THEOREM 6.0.1. *Let* $H_0, \ldots, H_N$ *be a set of hyperplanes in* P^n *in general position, and let* $\epsilon > 0$. *Then:*

(a). *(analytic case) If* $f \colon C \to P^n$ *is a holomorphic map whose image is not contained in any hyperplane, then*

$$\sum_{i=0}^{N} m(H_i, r) \leq (n + 1 + \epsilon)T(r) + O(1) \qquad //$$

(b). *(algebraic case) If* k *is a number field and* S *a finite set of places of* k, *then*

$$\sum_{i=0}^{N} m(H_i, P) \leq (n + 1 + \epsilon)h(P) + O(1)$$

for all P *outside a finite set of hyperplanes.*

For definitions, see Chapter 3. We note that (b) is merely (2.2.4).

The analytic case of the theorem was first proved by E. Cartan [**Car 2**], and later (using much longer proofs) by H. and J. Weyl and L. Ahlfors [**W-W**], [**A**]. The algebraic case was proved by Schmidt [**Schm 1**] in the late 1960's. His proof actually has much in common with the proofs of the Weyls and Ahlfors; the purpose of this chapter is to describe the similarities in detail. We follow Ahlfors' exposition most closely. For a more modern treatment see also Cowen and Griffiths [**Co-G**], although their version is not as close to Schmidt's. We also note that the Weyl-Ahlfors method has much in common with the approach used in [**Car 1**].

The similarity is not perfect, nor is it evident in all parts of the proof. It is most visible in the later parts of the proof, those parts dealing with the multidimensionality of $\mathbf{P}^n$. In that case, we still do not quite have a common proof, but the major steps in each case are the same. We will therefore assume the first halves of the proofs (the parts which roughly extend the proofs of Roth and Nevanlinna) and present only the second halves of the proofs. We do not intend for this to be a complete exposition of the proofs; for more details, see the above references.

For the sake of comparison, we have changed Schmidt's proof in three ways. First, we prove his result directly over number fields, instead of proving it over $\mathbf{Q}$ and then reducing the general case to this special case over $\mathbf{P}^{n[k:Q]}$. This brings the proof closer to Ahlfors' proof, since the latter had no such reduction. Unfortunately, it also makes the proof more difficult; in particular, we need a theory of successive minima over number fields. This is presented in Section 1.

Another change is technical and hard to explain. Instead of immediately proving a result on successive minima, we first prove a result (Theorem 6.4.3) which is quite close to a step in Ahlfors' proof (Theorem 6.5.1). This result then implies the successive minima result fairly easily (and vice versa). Section 4 contains the proof of this result, which has the same ingredients as Schmidt's proof, but combines them in a different way.

Finally, to simplify the exposition we assume the hyperplanes H_i are all defined over k (instead of over $\overline{\mathbf{Q}}$). As noted in Remark 2.2.3, this does not imply a loss of generality provided that all H_i and their conjugates over k all lie in general position.

In both proofs we will be using homogeneous coordinates in $\mathbf{A}^{n+1}$. In the analytic case, let $\mathbf{x} = \mathbf{x}(t)$ denote a vector of $n+1$ holomorphic functions describing the curve f. In the algebraic case, let vectors $\mathbf{x} \in O_{k,S}^{n+1}$ correspond to homogeneous coordinates of the points P.

§1. Successive Minima

Of all the steps in Schmidt's proof, the steps involving successive minima are most likely to cause problems when extending the proof to number fields. Certainly the geometrical intuition falls apart in the more general case; however, the algebraic formalisms can be extended so as to work in a more general proof. The purpose of this section, therefore, is to develop a theory of successive minima which is valid over number fields.

This goal, of course, is nothing new. In fact, Minkowski himself looked at successive minima for imaginary quadratic fields ([**Min**], [**R-SD**]). More recently, a number of authors have worked on successive minima over the adeles ([**Mac**], [**Bo-V**]). Our approach however, will use the number field k itself instead of its adeles. This is the approach used in [**R-SD**].

Before introducing this approach, however, it would be useful to recall the "usual" definitions and theorems of successive minima. Thus, let C be a convex, centrally symmetric body in $\mathbf{R}^n$ and let Λ be an n-dimensional lattice. Let $2^n V$ and $d(\Lambda)$ be their respective volumes. Then the <u>successive minima</u> of Λ relative to C are numbers $\lambda_1, \ldots, \lambda_n$ such that λ_j is the smallest real number for which $\lambda_j C \cap \Lambda$ has at least j linearly independent elements. For example, λ_1 is the smallest real number for which $\lambda_1 C \cap \Lambda \neq \{0\}$.

Then Minkowski's first theorem, which states that a symmetric convex body of volume 2^n has at least one nontrivial lattice point, merely states that

$$\lambda_1^n V \leq d(\Lambda).$$

It is Minkowski's stronger second theorem, however, which is central to the study of successive minima. It states,

$$\lambda_1 \lambda_2 \ldots \lambda_n V \leq d(\Lambda).$$

It is proved by stretching and shearing the body C to give a new convex body C' of volume $2^n V \lambda_2 \lambda_3 \ldots \lambda_n / \lambda_1^{n-1}$ and then applying the first theorem. For a complete proof, see [**Cas**].

The converse of the above inequality is trivial. Indeed, if $\mathbf{x}^{(1)}, \ldots, \mathbf{x}^{(n)}$ are n linearly independent points in Λ such that $\mathbf{x}^{(j)} \in \lambda_j C$ for all j, then the octahedron with vertices $\pm \mathbf{x}^{(j)}/\lambda_j$ must be contained in C. That octahedron has volume $2^n \det(x_i^{(j)})/n! \, \lambda_1 \lambda_2 \ldots \lambda_n \geq 2^n d(\Lambda)/n! \, \lambda_1 \lambda_2 \ldots \lambda_n$, so

$$\frac{d(\Lambda)}{n!} \leq \lambda_1 \lambda_2 \ldots \lambda_n V.$$

In Schmidt's original case, $\Lambda = \mathbf{Z}^n$ and C is a parallelopiped given by

(6.1.1) $$A_i |L_i(\mathbf{x})| \leq 1, \qquad 1 \leq i \leq n,$$

with $\prod A_i = 1$. This has volume $|\det L_i|^{-1}$, $\det L_i$ being the determinant of the coefficients appearing in L_i. Then the above inequalities could be summarized as,

(6.1.2) $$\frac{1}{n!} \leq \frac{\lambda_1 \lambda_2 \ldots \lambda_n}{|\det L_i|} \leq 1.$$

The remainder of this section will introduce a generalized form of (6.1.2) valid over number fields. Therefore, define a <u>length</u> <u>function</u> on $\mathbf{R}^n$ (or $\mathbf{Q}^n$) to be a continuous real-valued function $f(\mathbf{x})$ such that $f(\mathbf{x}) \geq 0$ for all $\mathbf{x}$, $f(\mathbf{x}) > 0$ for some $\mathbf{x}$, and $f(t\mathbf{x}) = |t| f(\mathbf{x})$. The convex body (6.1.1) could then be given as the region $f(\mathbf{x}) \leq 1$ if $f(\mathbf{x})$ is the length function $\max\limits_{1 \leq i \leq n} A_i |L_i(\mathbf{x})|$. In the number field case, we define the length function as follows. For each $v \in S$ let $L_{v,1}, L_{v,2}, \ldots, L_{v,n}$ be n linearly independent linear forms with coefficients in k, and let $A_{v,1}, \ldots, A_{v,n}$ be positive real numbers with

$$\prod_{i=1}^{n} A_{v,i} = 1$$

for each $v \in S$. Let $d = [k : \mathbf{Q}]$. Then let the length function be the following.

$$(6.1.3) \qquad f(\mathbf{x}) = \left[\prod_{v \in S} \max_{1 \leq i \leq n} A_{v,i} \| L_{v,i}(\mathbf{x}) \|_v \right]^{1/d}$$

This length function satisfies conditions roughly equivalent to those above: $f(\mathbf{x}) \geq 0$ for all $\mathbf{x}$, $f(\mathbf{x}) > 0$ for some $\mathbf{x}$, and $f(t\mathbf{x}) = N_S(t)^{1/d} f(\mathbf{x})$. (Here N_S is roughly the absolute value of the norm with a few added factors for finite S: $N_S(t) = \prod_{v \in S} \| t \|_v$. If t is a rational integer prime to S, then $f(t\mathbf{x}) = |t| f(\mathbf{x})$ as before.) In particular $\mathbf{x}$ can be scaled by a unit without changing the length function.

Relative to this length function, we can define the ith successive minimum of $O_{k,s}^n$ as the smallest real number λ such that $f(\mathbf{x}) \leq \lambda$ for at least i linearly independent points $\mathbf{x} \in O_{k,s}^n$. In the above, the words "linearly independent" mean linearly independent over k, so that there are n successive minima.

Having made the appropriate definitions, we now proceed to prove (6.1.2) in the number field case. Generally, the approach will be to use the classical case to prove the more general case for factors involving the infinite places, then to use extra arguments for the finite planes. A few more difficulties arise in this case, however, since the new distance function describes a star-body rather than a convex body.

For the proofs of both halves of (6.1.2), it will be useful to make a few definitions. Let $\mathbf{x}^{(1)}, \ldots, \mathbf{x}^{(n)}$ be a basis of k^n with S-integral coordinates such that $f(\mathbf{x}^{(i)}) = \lambda_1$. Since $f(\mathbf{x})$ is not affected when $\mathbf{x}$ is multiplied by an S-unit, we may assume the coordinates of $\mathbf{x}^{(i)}$ to be S-integers. Having fixed $\mathbf{x}^{(i)}$, define

$$\lambda_{v,j} = \max_{1 \leq i \leq n} A_{v,i} \| L_{v,i}(\mathbf{x}^{(j)}) \|_v$$

so that

(6.1.4)
$$\lambda_j = \sqrt[d]{\prod_{v \in S} \lambda_{v,j}}.$$

The general version of the left-hand side of (6.1.2) is then,

LEMMA 6.1.5.

$$\sqrt[d]{\prod_v \| \det x_i^{(j)} \|_v \cdot c_v} \leq \frac{\lambda_1 \lambda_2 \ldots \lambda_n}{\prod_v \| \det L_v \|_v^{1/d}}$$

where $c_v = 1/n!$ if v is real, $2^n(2n)!$ if complex, and 1 if non-archimedean.

PROOF: Using (6.1.4), the inequality can be proved by proving

(6.1.6)
$$\| \det x_i^{(j)} \|_v \cdot c_v \leq \frac{\prod_{j=1}^n \lambda_{v,j}}{\| \det L_v \|_v}$$

for each v and then taking the product over all $v \in S$.

First assume v is a real place, and let σ be the associated embedding of k into $\mathbf{R}$. Then the points,

(6.1.7)
$$\pm \frac{\sigma(\mathbf{x}^{(j)})}{\lambda_{v,j}}$$

lie in the symmetric convex body,

(6.1.8)
$$\max_{1 \leq i \leq n} A_{v,i} |L_{v,i}(\mathbf{x})| \leq 1.$$

(Here $L_{v,i}$ is regarded as a form over $\mathbf{R}^n$ by applying σ to its coefficients.) The convex body spanned by the points (6.1.7) lies inside the body (6.1.8). Since the volumes of these bodies are

$$\frac{2^n |\det \sigma(x_i^{(j)})|}{n! \prod_j \lambda_{v,j}} = \frac{2^n \| \det x_i^{(j)} \|_v}{n! \prod_j \lambda_{v,j}}$$

and $2^n \| \det L_v \|_v$, respectively, the inequality (6.1.6) follows immediately.

If v is complex, the inequality is proved in basically the same way. The points $w\sigma(\mathbf{x}^{(j)})/\sqrt{\lambda_{v,j}}$ lie in the convex body (6.1.8) if $|w| \leq 1$. They span a convex body of volume $(2\pi)^n \| \det x_i^{(j)} \|_v / (2n)! \prod_j \lambda_{v,j}$ which lies inside the body (6.1.8) having a volume of $\pi^n \| \det L_v \|_v$. Thus (6.1.6) holds if v is complex.

Finally, assume v is non-archimedean. Again, this is a volume computation in k_v using the volume of [**Mah**, Section 7]. The equivalent of (6.1.7) is now,

$$\left\{ a_1 \frac{\sigma(\mathbf{x}^{(1)})}{\pi^{f_1}} + \cdots + a_n \frac{\sigma(\mathbf{x}^{(n)})}{\pi^{f_n}} \mid a_1, \ldots, a_n \in \mathcal{O}_{k,v} \right\},$$

where π is a uniformizing parameter for k_v and f_j is an integer for which $\| \pi^{f_j} \|_v = \lambda_{v,j}$. By [**Mah**, Section 8], a linear transformation A changes volumes by a factor of $\| \det A \|_v$; therefore this body has volume $\| \det x_i^{(j)} \|_v \sqrt{\prod_j \lambda_{v,j}}$. It lies inside the body (6.1.8) which has volume $\| \det L_v \|_v$; thus (8) holds. $\square$

In the remainder of this section, we derive a generalization of the right-hand side of (6.1.2), namely:

$$(6.1.9) \qquad \frac{\lambda_1 \lambda_2 \ldots \lambda_n}{\sqrt[d]{\prod_v \| \det L_v \|_v}} \leq c_k.$$

In the case of ordinary integers $\mathcal{O}_k$, this is Theorem 3 of [**R-SD**]. The general case will be derived from [**Bo-V**], along with the derivation of C_k. The setup of [**Bo-V**] is somewhat different from ours; we recall it here. Let $S \subseteq \mathbf{A}_k^n$ be a subset of the n-fold product of adeles of k. This subset will be of the form,

$$S = S_\infty \times \prod_{v \notin S} S_v$$

where S_v is an $\mathcal{O}_v$-module of rank n and $S_v = \mathcal{O}_v^n$ for almost all v. S_∞ is a convex subset of $\mathbf{R}^{nr_1} \times \mathbf{C}^{nr_2}$ where r_1 and r_2 are the numbers of real and complex places of k, respectively; $r_1 + 2r_2 = d$. S is given a volume,

$$V(S) = V(S_\infty) \prod_{v \notin S} V(S_v).$$

The volume $V(S_\infty)$ is the usual Lebesgue measure on $\mathbf{R}^{nr_1} \times \mathbf{C}^{nr_2}$ multiplied by 2^{nr_2}; the volumes $V(S_v)$ are $N(\mathcal{D}_v^{-1})^{r/2}$ times the volumes defined earlier; $\mathcal{D}_v^{-1}$ is the inverse different.

We let $\mathbf{R}_{\geq 0}$ act on $\mathbf{A}_k^d$ by dilation at the infinite places; it leaves the finite places alone. The successive minima relative to this action are:

$$\mu_i = \min \{\lambda > 0 \mid \lambda S \cap k^n \text{ contains at least } i$$
$$\text{linearly independent vectors}\}.$$

THEOREM 6.1.10 [Bo-V]. *With the above assumptions and definitions,*

$$(\mu_1 \mu_2 \ldots \mu_n)^d V(S) \geq 2^{nd}.$$

Returning to our situation, we let $S_v = O_v^n$ if $v \in S$;

$$S_v = \{\mathbf{x} \in O_v^n \mid A_{v,i} \|L_{v,i}(\mathbf{x})\|_v \leq 1\} \qquad \text{if } v \in S \setminus S_\infty;$$
$$S_\infty = \{\mathbf{x} \in \mathbf{R}^{r_1} \times \mathbf{C}^{r_2} \mid \sum_v N_v \max_i A_{v,i} |L_{v,i}^{\sigma_v}(\mathbf{x}_v)| \leq d\}$$

where $N_v = 1$ if real or 2 if complex, and $L_{v,i}^{\sigma_v}$ is the form obtained by applying the injection $\sigma_v \colon k \to \mathbf{C}$ to the coefficients of $L_{v,i}$. This defines a convex subset of adelic n-space; it remains only to compare the two notions of successive minima and volumes. First, we know that S_∞ is contained in the star-body

$$\{x \in \mathbf{R}^{r_1} \times \mathbf{C}^{r_2} \mid \prod_v \max_i A_{v,i} \|L_{v,i}^{\sigma_v}(\mathbf{x}_v)\|_v \leq 1\}.$$

Thus, if $\mathbf{x}^{(1)}, \ldots, \mathbf{x}^{(n)}$ are linearly independent points in k^n for which $x^{(i)} \in \mu_i S$, then the length function satisfies

$$f(\mathbf{x}^{(i)}) \leq \mu_i.$$

Thus

$$\lambda_i \leq \mu_i.$$

Let $C(r_1, r_2, n)$ be the volume of the solid in $\mathbf{R}^{nr_1} \times \mathbf{C}^{nr_2}$ given by,

$$\{\mathbf{x} = (x_{v,i}), v \in S_\infty, 1 \leq i \leq n \mid \sum_v N_v \max |x_{v,i}| \leq d\}.$$

Then the volume $V(S)$ in the sense described above is,

$$V(S) = \prod_{v \in S} \| \det L_{v,i} \|_v^{-1} \cdot |D_k|^{-n/2} \cdot 2^{nr_2} \cdot C(r_1, r_2, n).$$

Thus Theorem 6.1.10 implies,

$$\lambda_1 \lambda_2 \ldots \lambda_n \leq \left[\frac{2^{n(r_1+r_2)} |D_k|^{n/2} \prod_{v \in S} \| \det L_{v,i} \|_v}{C(r_1, r_2, n)} \right]^{1/d}.$$

Combining this with Lemma 6.1.5 gives,

THEOREM 6.1.11. *Let* $\lambda_1,\ldots,\lambda_n$ *denote the successive minima of* $O_{k,S}^n$ *with respect to the length function (6.1.3). Then*

$$\left(\frac{1}{n!}\right)^{r_1}\left(\frac{2^n}{(2n)!}\right)^{r_2} \leq \frac{(\lambda_1\lambda_2\ldots\lambda_n)^d}{\prod_{v\in S}\|\det L_{v,i}\|_v} \leq \frac{2^{n(r_1+r_2)}|D_k|^{n/2}}{C(r_1,r_2,n)}.$$

As an example of how the right-hand bound behaves as the number field becomes large, a fair amount of computation will show that,

$$C(0,r_2,2) = \frac{(24\pi^2 r_2^4)^{r_2}}{(4r_2)!}.$$

This covers binary linear forms over totally complex fields.

REMARK 6.1.12. Theorem 6.1.11 holds not only for length functions defined by linear forms, but also for more general length functions of the form,

$$f(\mathbf{x}) = \prod_{v\in S} f_v(\mathbf{x}),$$

where $f_v(\mathbf{x})$ is a convex length function on k_v^n.

§2. Davenport's Lemma

We keep the notations of the last section. In short, these notations are: for each $v \in S$, let $L_{v,1},\ldots,L_{v,n}$ be n linearly independent linear forms on k^n with coefficients in k. Moreover, assume that the collection of all $L_{v,i}$, except for duplicates, lies in general position. This slightly weakens the proof, but it will keep the proof both simpler and closer to its analytic equivalent.

For all v and i as above, we have positive real constants $A_{v,i}$ with,

$$\prod_{i=1}^{n} A_{v,i} = 1 \qquad \text{for all } v \in S.$$

Let the height of $A = (A_{v,i})$ be,

$$H(A) = \prod_{v\in S} \max(A_{v,1},\ldots,A_{v,n}).$$

Finally, let $\Pi(A)$ denote the star-body with length function,

$$f(\mathbf{x})^{[k:\mathbb{Q}]} = \prod_{v\in S} \max_{0\leq i\leq n} A_{v,i}\|L_{v,i}(\mathbf{x})\|_v,$$

and let $\lambda_1,\ldots,\lambda_n$ be the successive minima of $\Pi(A)$. Davenport's lemma then states:

LEMMA 6.2.1. *With the above notations, assume that* $\rho_1, \ldots, \rho_n$ *are positive real numbers with,*

(6.2.2)
$$\rho_1 \geq \rho_2 \geq \cdots \geq \rho_n;$$

(6.2.3)
$$\rho_1 \lambda_1 \leq \rho_2 \lambda_2 \leq \cdots \leq \rho_n \lambda_n;$$

(6.2.4)
$$\rho_1 \rho_2 \cdots \rho_n = 1.$$

Then for each $v \in S$ *and each* $1 \leq i \leq n$ *there exists a real constant* $\rho_{v,i}$ *and constants* c_1, c_2 *depending only on* k, S, *such that the successive minima* $\hat{\lambda}_i$ *of the length function,*

$$\hat{f}(\mathbf{x}) = \prod_{v \in S_\infty} \max_i \rho_{v,i} A_{v,i} \| L_{v,i}(\mathbf{x}) \|_v$$

satisfy

(6.2.5)
$$c_1 \lambda_1 \rho_i \leq \hat{\lambda}_i \leq c_2 \lambda_i \rho_i.$$

Also

(6.2.6)
$$\prod_{i=0}^{n} \rho_{v,i} = 1 \qquad \text{for each } v \in S.$$

Finally, letting $N_v = 2$ *if* v *is complex,* 1 *if* v *is real, or* 0 *if* v *is non-archimedean, we have,*

(6.2.7)
$$\max_{1 \leq i \leq n} \rho_{v,i} = \rho_1^{N_v} \qquad \text{for each } v \in S.$$

PROOF: For convenience of notation, assume $A_{v,i} = 1$ for all v and i. For each $1 \leq j \leq n$ let $\mathbf{x}^{(j)}$ be a vector in $O_{k,S}^n$ such that

$$f(\mathbf{x}^{(j)}) = \lambda_j.$$

Scaling each $\mathbf{x}^{(j)}$ by some unit, we may assume that,

(6.2.8)
$$\max_i \| L_{v,i}(\mathbf{x}^{(j)}) \|_v \gg\ll \lambda_j^{N_v}.$$

The constant c_3 in $\gg\ll$ depends only on k and S.

For $1 \leq j \leq n$, let E^j be the subspace of k^n spanned by $\mathbf{x}^{(1)}, \ldots, \mathbf{x}^{(j)}$. For each $v \in S_\infty$ the n linear forms $L_{v,i}$ satisfy a linear relation,

$$\alpha_{v,1} L_{v,1} + \cdots + \alpha_{v,n} L_{v,n} \equiv 0 \qquad \text{on } E^{n-1}.$$

Reorder the $L_{v,i}$ such that $\|\alpha_{v,n}\|_v$ is the largest of the $\|\alpha_{v,i}\|_v$. Hence, if $\mathbf{x} \in E^{n-1}$, then

$$\alpha_{v,n} L_{v,n}(\mathbf{x}) = -\alpha_{v,1} L_{v,1}(\mathbf{x}) - \cdots - \alpha_{v,n-1} L_{v,n-1}(\mathbf{x});$$
$$\|L_{v,n}(\mathbf{x})\|_v \leq \|L_{v,1}(\mathbf{x})\|_v + \cdots + \|L_{v,n-1}(\mathbf{x})\|_v$$

since $\|\alpha_{v,i}/\alpha_{v,n}\|_v \leq 1$. Thus for $\mathbf{x} \in E^{n-1}$,

$$\|L_{v,1}(\mathbf{x})\|_v + \cdots + \|L_{v,n-1}(\mathbf{x})\|_v \geq \frac{1}{2}(\|L_{v,1}(\mathbf{x})\|_v + \cdots + \|L_{v,n}(\mathbf{x})\|_v).$$

By induction, reorder the remaining $L_{v,i}$ so that,

$$(6.2.9) \quad \|L_{v,1}(x)\|_v + \cdots + \|L_{v,i}(x)\|_v \geq \frac{1}{2^{n-i}} \|L_{v,1}(x)\|_v + \cdots + \|L_{v,n}(x)\|_v$$

on E^i. Having permuted the $L_{v,i}$ in this manner, let $\rho_{v,i} = \rho_i^{N_v}$. These choices automatically satisfy (6.2.6) and (6.2.7). Now assume $\mathbf{x} \notin E^i$. Then, for some j, $j > i$, $\mathbf{x} \notin E^{j-1}$ but $\mathbf{x} \in E^j$. Hence,

$$\max(\rho_{v,1}\|L_{v,1}(\mathbf{x})\|_v, \ldots, \rho_{v,n}\|L_{v,n}(\mathbf{x})\|_v)$$
$$\geq \max(\rho_{v,1}\|L_{v,1}(\mathbf{x})\|_v, \ldots, \rho_{v,j}\|L_{v,j}(\mathbf{x})\|_v)$$
$$\geq \rho_{v,j} \max(\|L_{v,1}(\mathbf{x})\|_v, \ldots, \|L_{v,j}(\mathbf{x})\|_v)$$
$$\text{(by 6.2.2)}$$
$$\geq \frac{\rho_{v,j}}{j}(\|L_{v,1}(\mathbf{x})\|_v + \cdots + \|L_{v,j}(\mathbf{x})\|_v)$$
$$\geq \frac{\rho_{v,j}}{j \cdot 2^{n-j}}(\|L_{v,1}(\mathbf{x})\|_v + \cdots + \|L_{v,n}(\mathbf{x})\|_v)$$
$$\text{(by 6.2.9)}$$
$$\geq \frac{\rho_{v,j}}{2^n} \max(\|L_{v,1}(\mathbf{x})\|_v, \ldots, \|L_{v,n}(\mathbf{x})\|_v)$$
$$\geq 2^{-n}(\rho_j \lambda_j)^{N_v}/c_3 \qquad \text{(by 6.2.8)}.$$

Thus

$$\hat{f}(\mathbf{x}) \geq c_1 \rho_j \lambda_j$$
$$\geq c_1 \rho_{i+1} \lambda_{i+1} \qquad \text{(by 6.2.3)}.$$

This holds for all $\mathbf{x} \notin E^i$, proving the first half of (6.2.5). The second half follows from (6.2.4) and Theorem 6.1.11. $\qquad\qquad\square$

§3. Multilinear Algebra

In this section we briefly introduce Grassman algebras on $V = F^n$, where F is a field. Of particular interest is the way in which wedge products affect successive minima in Schmidt's case, and how they affect the associated curves of a meromorphic curve, in Ahlfors' case. The appropriate references for this section are [**Schm 1**] and [**Greub**].

Let V^* denote the dual space of V. If $\mathbf{y} \in V^*$ and $\mathbf{x} \in V$, then we write,

$$(\mathbf{x} \cdot \mathbf{y}) = \mathbf{y}(\mathbf{x});$$

For notational convenience also let,

$$|\mathbf{x} \cdot \mathbf{y}| = |(\mathbf{x} \cdot \mathbf{y})|.$$

Let $\mathbf{e}_1, \ldots, \mathbf{e}_n$ be a basis of V and $\mathbf{e}_1^*, \ldots, \mathbf{e}_n^*$ be the corresponding dual basis of V^*. If $\mathbf{x} = \sum x_i \mathbf{e}_i$ and $\mathbf{y} = \sum y_i \mathbf{e}_1^*$, then

$$\mathbf{x} \cdot \mathbf{y} = \sum x_i y_i.$$

Note that y_i is not conjugated; this is because sometimes we will want F to be a p-adic field. Consequently, we will have no notion of orthogonality on V, and we will always carefully distinguish V from V^*. For the lengths of vectors, we use the sup norm,

$$\|\mathbf{x}\| = \max \|x_i\|,$$

where x_i are the coordinates of $\mathbf{x}$ relative to some given basis.

DEFINITION 6.3.1. *Let* $0 \le p \le n$. *Then the space of p-vectors, denoted* $\wedge^p V$, *is the vector space over* F,

$$\wedge^p V = \big(\overbrace{F^n \otimes \cdots \otimes F^n}^{p \text{ times}} \big) / M,$$

where M *is the submodule generated by relations of the form,*

$$\mathbf{x}^{(1)} \otimes \cdots \otimes \mathbf{x}^{(i)} \otimes \mathbf{x}^{(i+1)} \otimes \cdots \otimes \mathbf{x}^{(p)} = -\mathbf{x}^{(1)} \otimes \cdots \otimes \mathbf{x}^{(i+1)} \otimes \mathbf{x}^{(i)} \otimes \cdots \otimes \mathbf{x}^{(p)}.$$

Let $\mathbf{x}^{(1)} \wedge \cdots \wedge \mathbf{x}^{(p)}$ denote the image of a vector $\mathbf{x}^{(1)} \otimes \cdots \otimes \mathbf{x}^{(p)}$ in $\wedge^p V$. A p-vector which can be written in this form is called <u>decomposable</u>. Any p-vector can be written as a finite sum of decomposable p-vectors.

Let $\mathbf{e}_1 = (1, 0, 0, \dots), \dots, \mathbf{e}_n = (0, 0, \dots, 1)$ be the standard basis for F^n. Then, as σ varies over all p-tuples $1 \le \sigma(1) < \sigma(2) < \cdots < \sigma(p) \le n$ the vectors

$$(6.3.2) \qquad E_\sigma = \mathbf{e}_{\sigma(1)} \wedge \cdots \wedge \mathbf{e}_{\sigma(p)} \in \wedge^p V$$

form a basis for $\wedge^p V$. We have $(\wedge^p V)^* \cong \wedge^p (V^*)$; its basis

$$E_\sigma^* = e_{\sigma(1)}^* \wedge \cdots \wedge \mathbf{e}_{\sigma(p)}^*$$

is dual to E_σ and defines the dot product by the usual formula.

EXAMPLES 6.3.3.

   (a). $\wedge^0 F^n \cong F$ canonically

   (b). $\wedge^1 F^n \cong F^n$ canonically

   (c). $\wedge^p F^n$ has dimension $\binom{n}{p}$. In particular, $\wedge^n F^n$ is (non-canonically) isomorphic to F.

   (d). $\wedge^p F^n$ is dual to $\wedge^{n-p} F^n$ by the pairing

$$\wedge \colon (\wedge^p F^n) \times (\wedge^{n-p} F^n) \to \wedge^n F^n \cong F.$$

   Thus $\wedge^{n-1} F^n \cong (F^n)^*$.

LEMMA 6.3.4 (LAPLACE IDENTITY). *For decomposable p-vectors X^p, Y^p, the dot product can be written as,*

$$\left(x^{(1)} \wedge \cdots \wedge x^{(p)} \right) \cdot \left(y^{(1)} \wedge \cdots \wedge y^{(p)} \right) = \det\left(x^{(i)} \cdot y^{(j)} \right),$$

where the dot in the right hand side is the dot product in F^n.

PROOF: By linearity, this reduces to the case where all $x^{(i)}$ and $y^{(j)}$ are basis vectors $\mathbf{e}_{i_i}$ and $e_{j_j}^*$ in V and V^*, respectively; then it is trivial. $\square$

COROLLARY 6.3.5. *With respect to the standard basis $\mathbf{e}_\sigma$, the coordinates x_σ of $X = \mathbf{x}^{(1)} \wedge \cdots \wedge \mathbf{x}^{(p)}$ are $p \times p$ determinants,*

$$x_\sigma = \det\left(x_{\sigma(j)}^{(i)} \right)$$

where $\mathbf{x}^{(i)} = \left(x_1^{(i)}, \dots, x_n^{(i)} \right)$.

COROLLARY 6.3.6. $\mathbf{x}^{(1)} \wedge \cdots \wedge \mathbf{x}^{(p)} \ne 0$ *if and only if $\mathbf{x}^{(1)}, \dots, \mathbf{x}^{(p)}$ are linearly independent. If so, then $X^p = \mathbf{x}^{(1)} \wedge \cdots \wedge \mathbf{x}^{(p)}$ is proportional to $Y^p = \mathbf{y}^{(1)} \wedge \cdots \wedge \mathbf{y}^{(p)}$ if and only if the p-dimensional subspaces $S_x = \langle \mathbf{x}^{(1)}, \dots, \mathbf{x}^{(p)} \rangle$ and $S_y = \langle \mathbf{y}^{(1)}, \dots, \mathbf{y}^{(p)} \rangle$ coincide.*

PROOF: The first assertion follows from 6.3.5. Now assume that $\mathbf{x}^{(1)}, \dots, \mathbf{x}^{(p)}$ are linearly independent.

If $\mathbf{y}^{(1)},\ldots,\mathbf{y}^{(p)}$ also span $S = S_x$, then $\mathbf{y}^{(1)} \wedge \cdots \wedge \mathbf{y}^{(p)}$ is a nonzero element of

$$\wedge^p S \subseteq \wedge^p V.$$

But $\wedge^p S$ is a one dimensional space, so X^p and Y^p must be proportional. Conversely, if X^p and Y^p are proportional, then

$$S_x = \{\mathbf{x} \in V \mid \mathbf{x} \wedge X^p = 0\},$$

so that $S_x = S_y$. $\qquad\qquad\square$

Thus $G(p, F^n)$, the Grassmannian space of p-dimensional subspaces of F^n, is a subset of $\wedge^p F^n$. The coordinates on $\wedge^p F^n$ give the Plücker coordinates on $G(p, F^n)$.

In Schmidt's situation, we will have n linearly independent linear forms $L_1,\ldots,L_n$, and we will want to define $\binom{n}{p}$ independent linear forms on $\wedge^p F^n$. Let $\mathbf{b}_1,\ldots,\mathbf{b}_n$ be vectors in V^* such that

$$L_i(\mathbf{x}) = \mathbf{x} \cdot \mathbf{b}_i.$$

For $\sigma = (\sigma(1),\ldots,\sigma(p))$ as before, let

$$B_\sigma = \mathbf{b}_{\sigma(1)} \wedge \cdots \wedge \mathbf{b}_{\sigma(p)}.$$

These vectors define linear forms $L_\sigma(X)$ on $\wedge^p F^n$, which are linearly independent. In fact,

LEMMA 6.3.7. *Let* $\det(L_i)$ *denote the determinant of the matrix of the coordinates of the vectors* $\mathbf{b}_i$. *Then*

$$\det(L_\sigma) = \det(L_i)^{\binom{n-1}{p-1}}.$$

PROOF: See [**Schm 1**, Ch. IV, Lemma 6E]. $\qquad\qquad\square$

We now let $F = k$ be a number field, and let S, $L_{v,i}$, $A_{v,i}$, and $\lambda_1,\ldots,\lambda_n$ be as in Section 2. As with the linear forms, define

(6.3.8)
$$A_{v,\sigma} = A_{v,\sigma(1)} A_{v,\sigma(2)} \cdots A_{v,\sigma(p)}$$
$$\lambda_\sigma = \lambda_{\sigma(1)} \lambda_{\sigma(2)} \cdots \lambda_{\sigma(p)}.$$

LEMMA 6.3.9. *With* $L_{v,\sigma}$ *and* $A_{v,\sigma}$ *as above, and* $X = \mathbf{x}^{(1)} \wedge \cdots \wedge \mathbf{x}^{(p)}$, *we have,*

$$A_{v,\sigma} \|L_{v,\sigma}(X)\|_v \leq c_v \prod_{i=1}^{P} \max_{1 \leq j \leq p} A_{v,\sigma(j)} \|L_{v,\sigma(j)}(\mathbf{x}^{(i)})\|_v,$$

where $c_v = n!$ *if* v *is real,* $(n!)^2$ *if complex, and* 1 *if non-archimedean.*

PROOF: Immediate from the Laplace identity. $\qquad\qquad\square$

PROPOSITION 6.3.10. *Order the σ's so that*

$$\lambda_{\sigma_1} \leq \cdots \leq \lambda_{\sigma_N},$$

where $N = \binom{n}{p}$. Let $\mu_1, \ldots, \mu_N$ be the successive minima of the system $(A_{v,\sigma}, L_{v,\sigma})$. Then

(6.3.11) $\lambda_{\sigma_i} \ll \mu_i \ll \lambda_{\sigma_i}$ *for all i.*

Moreover, $\sigma_1 = (1, 2, \ldots, p)$ and $\sigma_2 = (1, 2, \ldots, p-1, p+1)$.

PROOF: Let f be the length function on k^n defined by $L_{v,i}$ and $A_{v,i}$; let F be the length function on $\wedge^p k^n$ defined by the corresponding $L_{v,\sigma}$, $A_{v,\sigma}$. Let $\mathcal{O}$ denote the ring of S-integers, and let $\mathbf{x}^{(1)}, \ldots, \mathbf{x}^{(n)}$ be linearly independent vectors in $\mathcal{O}^n$ with

$$f(\mathbf{x}^{(i)}) = \lambda_i.$$

Define p-vectors $X^{(\sigma)} = \mathbf{x}^{(\sigma(1))} \wedge \cdots \wedge \mathbf{x}^{(\sigma(p))}$. Then

$$F(X^{(r)}) \ll \prod_{i=1}^{P} f(\mathbf{x}^{(\sigma(i))})$$

by Lemma 6.3.9. Since the vectors $X^{(\sigma)}$ are linearly independent this gives the right-hand half of (6.3.11). The other half follows from (6.3.7), (6.1.11), and the fact that

$$\pi_\sigma \lambda_\sigma = (\pi_i \lambda_i)^{\binom{n-1}{p-1}}.$$

The final assertion in this proposition is trivial. $\square$

   Let $Y^q \in \wedge^q V^*$ if $p > q$ then let the interior product $(X^p \cdot Y^q)$ be the $(p-q)$-vector such that, for all dual $(p-q)$-vectors Z,

(6.3.12) $(X^p \cdot Y^q) \cdot Z = (X^p \cdot (Y^q \wedge Z)).$

Symmetrically, we can define $(X^p \cdot Y^q)$ if $p < q$. For simplicity of notation, write $|X^p \cdot Y^q| = |(X^p \cdot Y^q)|$. If E is a subspace of dimension q, then let E^q be any vector of unit length associated to this subspace, as in 6.3.6. This is unique up to multiplication by a scalar of absolute value one. We use the notation of q-dimensional subspaces and their associated unit q-vectors interchangeably.

Assuming we have an isomorphism $V \xrightarrow{\sim} V^*$, i. e. a notion of orthogonality, these symbols can be interpreted as follows. If $p \leq q$, then $|X^p \cdot E^q|$ is the length of the projection of X^p onto E^q. If $p \geq q$, then a corresponding interpretation follows by replacing X^p and E^q by multivectors of the same lengths corresponding to the orthogonal complements. These facts can be seen by choosing coordinates so that E^q is spanned by the first q coordinate vectors.

LEMMA 6.3.13. *Let $m \geq 1$ be an integer and let $\mathbf{x}_0, \ldots, \mathbf{x}_m \in V$; $\mathbf{y}_0, \ldots, \mathbf{y}_m \in V^*$. Then*

$$((\mathbf{x}_0 \wedge \mathbf{x}_1) \wedge (\mathbf{x}_0 \wedge \mathbf{x}_2) \wedge \cdots \wedge (\mathbf{x}_0 \wedge \mathbf{x}_m)) \cdot ((\mathbf{y}_0 \wedge \mathbf{y}_1) \wedge \cdots \wedge (\mathbf{y}_0 \wedge \mathbf{y}_m))$$
$$= (\mathbf{x}_0 \cdot \mathbf{y}_0)^{m-1}((\mathbf{x}_0 \wedge \cdots \wedge \mathbf{x}_m) \cdot (\mathbf{y}_0 \wedge \cdots \wedge \mathbf{y}_m)),$$

where the dot product on the left-hand side is in $\wedge^m(\wedge^2 V)$.

PROOF: If $\mathbf{x}_0, \ldots, \mathbf{x}_m$ are linearly dependent, then both sides are zero; therefore we may assume $\mathbf{x}_0, \ldots, \mathbf{x}_m$ form part of a basis for V. Let $\mathbf{x}_0^*, \ldots, \mathbf{x}_n^*$ be the dual basis for V^*, and let $y_{i0}, \ldots, y_{in}$ be the coordinates of $\mathbf{y}_i$ relative to this basis. Then we want to compare the $(01 \ldots m)$ coordinate of $\mathbf{y}_0 \wedge \cdots \wedge \mathbf{y}_m$ with the $((01)(02) \ldots (0m))$ coordinate of $((\mathbf{y}_0 \wedge \mathbf{y}_1) \wedge \cdots \wedge (\mathbf{y}_0 \wedge \mathbf{y}_m))$. These coordinates are, respectively,

$$\det(y_{ij})_{0 \leq i,j \leq m}$$

and

$$\det((y_{00} y_{ij} - y_{i0} y_{0j}))_{1 \leq i,j \leq m}.$$

This latter coordinate equals $y_{00}^{m-1} \det(y_{ij})$, as can be shown using properties of determinants. $\qquad \square$

LEMMA 6.3.14. *Let $X^{p-1} = \mathbf{x}^{(0)} \wedge \cdots \wedge \mathbf{x}^{(p-2)} \in \wedge^{p-1} V$. Let $\mathbf{x}, \mathbf{x}' \in V$. Let $E^{p-1} \in \wedge^{p-1} V^*$ and $\mathbf{y} = ((X^{p-1} \wedge \mathbf{x}) \cdot E^{p-1})$, $\mathbf{y}' = ((X^{p-1} \wedge \mathbf{x}') \cdot E^{p-1})$. Then*

$$\mathbf{y} \wedge \mathbf{y}' = (X^{p-1} \cdot E^{p-1})((X^{p-1} \wedge \mathbf{x} \wedge \mathbf{x}') \cdot E^{p-1}).$$

PROOF: (See also [Car 1, Ch. III, §2].) This reduces to the following identity of determinants. Let M be a $(p-1) \times (p-1)$ matrix; A and A', $(p-1) \times 1$ matrices; B and B', $1 \times (p-1)$ matrices; and c, d, e, f, scalars. Then

$$\left\| \begin{array}{cc} \begin{vmatrix} M & A \\ B & c \end{vmatrix} & \begin{vmatrix} M & A' \\ B' & d \end{vmatrix} \\ \begin{vmatrix} M & A \\ B' & e \end{vmatrix} & \begin{vmatrix} M & A' \\ B' & f \end{vmatrix} \end{array} \right\| = |M| \begin{vmatrix} M & A & A' \\ B & c & d \\ B' & e & f \end{vmatrix}$$

The left-hand side is the determinant of a 2×2 matrix whose elements are determinants of $p \times p$ matrices.

If this is true, then it holds as an identity of polynomials in the matrix elements, so it will suffice to prove it when M is nonsingular, which holds generically. Furthermore, if one applies the same row operators simultaneously to M, A, and A', then none of the above determinants changes. Therefore, we may assume that M is diagonal; in that case the identity is easily checked. $\qquad\square$

§4. The Start of the Algebraic Proof

In this section and Section 6 we will prove the algebraic version of Theorem 6.0.1, using the tools introduced in the previous three sections. To each point $P \in \mathbf{P}^n$ we assign homogeneous coordinates, obtaining a vector $x \in O_{k,S}^{n+1}$. Therefore (changing notation) let $V = k^{n+1}$. Our starting point is the following.

THEOREM 6.4.1. *Let $L_{v,i}$ be fixed linear forms on k^{n+1}, as in Section 2. Let $C_{v,i}$ be a collection of real constants such that $\sum_i C_{v,i} = 0$ for each $v \in S$. Let $\epsilon \in 0$. Let*

$$\overline{H}(x) = \prod_{v \in S} \|x\|_v$$

denote the height of a vector $x \in O_{k,S}^{n+1}$. Then there exists a finite set $S \subseteq V$ such that if $x \in O_{k,S}^{n+1}$ is a vector and $\mathbf{w}_1,\ldots,\mathbf{w}_n \in (O_{k,S}^{n+1})^*$ is a basis for the subspace of V^* which is zero on x, and if

$$\|L_{v,i}(w_j)\|_v \leq \overline{H}(x)^{C_{v,i}-\epsilon} \qquad \text{for all } v \in S, \; 0 \leq 1 \leq n, \; 1 \leq j \leq n,$$

then $x \in S$.

If $O_{k,S} = \mathbf{Z}$ then this is a consequence of Theorems 9A and 10B of Chapter VI of [Schm 1]. The general case can be proved by essentially the same proof, but we omit the details.

THEOREM 6.4.2. *Let ϵ and $L_{v,i}$ be as above. Let $c > 0$ be constant. Then there exists a finite set $S \subseteq V$ such that if $x \in O_{k,S}^{n+1}$ and $\mathbf{w}_1,\ldots,\mathbf{w}_n \in (O_{k,S}^{n+1})^*$ are as above, such that,*

$$\|L_{v,i}(\mathbf{w}_j)\|_v \leq \overline{H}(x)^c \qquad \text{for all } i, \; j, \; v;$$

and if

$$\prod_{v \in S} \prod_{i=0}^{n} \max_{1 \leq j \leq n} \|L_{v,i}(\mathbf{w}_j)\|_v \leq \overline{H}(x)^{-\epsilon}.$$

then $\mathbf{x} \in S$.

PROOF: This follows from Theorem 6.4.1 by a compactness argument.

$\square$

The purpose of this section is to prove the following theorem, which will be the starting point of the algebraic part of the proof.

THEOREM 6.4.3. *Let* $\mathbf{b}_0, \ldots, \mathbf{b}_N$ *be a set of vectors in* V^* *in general position. Let* $\epsilon > 0$. *Then there exists a finite set* S *of* V *such that if* $\mathbf{x} \in O_{k,S}^{n+1}$ *is not a scalar multiple of some vector in* S, *then there exists an* $\mathbf{x}' \in O_{k,S}^{n+1}$ *such that* $\mathbf{x} \wedge \mathbf{x}' \neq 0$ *and*

$$\log \sum_{v \in S} \frac{\|(\mathbf{x} \wedge \mathbf{x}') \cdot \mathbf{b}_i\|_v}{\|\mathbf{x}\|_v \|\mathbf{x} \cdot \mathbf{b}_i\|_v} < \epsilon \log \overline{H}(\mathbf{x})$$

for all $i = 0, \ldots, N$ *for which* $\mathbf{x} \cdot \mathbf{b}_i \neq 0$. *If* $\mathbf{x} \cdot \mathbf{b}_i = 0$ *then* $\mathbf{x}' \cdot \mathbf{b}_i = 0$.

It will suffice to prove the theorem under the assumption that $\mathbf{x} \cdot \mathbf{b}_i \neq 0$ for all i. Indeed, if $\mathbf{x} \cdot b_i = 0$ for some i, then we can reduce to the subspace perpendicular to $\mathbf{b}_i$ and use induction on n. Before proceeding with the proof of the theorem, we first need a lemma.

LEMMA 6.4.4. *Let* k *be a field with absolute value* $|\cdot|$ *and let* $V = k^{n+1}$. *Let* $\mathbf{b}_0, \ldots, \mathbf{b}_N$ *be* $N+1$ *vectors in* V^* *in general position. If for some index* i,

(6.4.5) $$|(\mathbf{x} \wedge \mathbf{x}') \cdot \mathbf{b}_i| > A|\mathbf{x}||\mathbf{x} \cdot \mathbf{b}_i|,$$

then there exists a constant $c > 0$ *depending only on the* $\mathbf{b}$ *'s, an index* ℓ *depending only on* $\mathbf{x}$ *and the* $\mathbf{b}$ *'s, and an index* j *depending also on* $\mathbf{x}'$, *such that,*

$$|(\mathbf{x} \wedge \mathbf{x}') \cdot (\mathbf{b}_j \wedge \mathbf{b}_\ell)| > cA|\mathbf{x}||\mathbf{x} \cdot \mathbf{b}_i|.$$

PROOF: Order the vectors $\mathbf{b}_i$ so that,

(6.4.6) $$|\mathbf{x} \cdot \mathbf{b}_0| \leq \cdots \leq |\mathbf{x} \cdot \mathbf{b}_N|.$$

Then $\mathbf{b}_0, \ldots, \mathbf{b}_n$ is a basis for V^*; let $\mathbf{b}_0^*, \ldots, \mathbf{b}_n^*$ be the corresponding dual basis. Then for any vector $\mathbf{y} \in V$ with coordinates $y_0, \ldots, y_n$ relative to $\mathbf{b}_0^*, \ldots, \mathbf{b}_n^*$,

$$|\mathbf{y}| \gg\ll \max(|y_0|, \ldots, |y_n|).$$

(Here, and in the sequel, all constants implicit in $\ll$, etc. will depend only on the $\mathbf{b}$'s.) Also, $|\mathbf{x}| \gg\ll |x_n| = |\mathbf{x} \cdot \mathbf{b}_n|$. Since the lemma is unchanged if we replace $\mathbf{x}'$ by $\mathbf{x}' - c\mathbf{x}$ for $c \in k$, we may assume $\mathbf{x}' \cdot \mathbf{b}_n = 0$.

We first claim that, with this choice of $\mathbf{x}'$, (6.4.5) implies,

$$\max_{0 \le j \le n} \frac{|\mathbf{x}' \cdot \mathbf{b}_j|}{|\mathbf{x} \cdot \mathbf{b}_j|} \gg A.$$

Indeed, just for the proof of this claim, change bases on V again so that the basis is $\{\mathbf{b}_0^*, \ldots, \mathbf{b}_{n-1}^*, \mathbf{u}\}$ where $\mathbf{u}$ is a unit vector proportional to $\mathbf{x}$. By (6.4.6) both this base change and its inverse have bounded coefficients, so that the sup norm with respect to this basis is $\gg\ll$ the length of a vector in V (i. e. the sup norm with respect to the standard basis of V).

Relative to this basis, the coordinates of $((\mathbf{x} \wedge \mathbf{x}') \cdot \mathbf{b}_i)$ are,

$$((\mathbf{x} \wedge \mathbf{x}') \cdot \mathbf{b}_i)_j = \begin{vmatrix} \mathbf{x} \cdot \mathbf{b}_i & \mathbf{x}' \cdot \mathbf{b}_i \\ x_j & x_j' \end{vmatrix}$$
$$= \begin{cases} (\mathbf{x} \cdot \mathbf{b}_i) x_j' & 0 \le j < n; \\ (\mathbf{x}' \cdot \mathbf{b}_i)(x_n), & j = n. \end{cases}$$

The length of this vector is then,

$$|(\mathbf{x} \wedge \mathbf{x}') \cdot \mathbf{b}_i| \gg\ll \max(|\mathbf{x}||\mathbf{x}' \cdot \mathbf{b}_i|, |\mathbf{x}'||\mathbf{x} \cdot \mathbf{b}_i|);$$

so that

$$\max\left(\frac{|\mathbf{x}' \cdot \mathbf{b}_i|}{|\mathbf{x} \cdot \mathbf{b}_i|}, \frac{|\mathbf{x}'|}{|\mathbf{x}|}\right) \gg A.$$

If the first term in the above max is the larger, then the claim is proved; otherwise, we know that

$$|\mathbf{x}' \cdot \mathbf{b}_j| \gg |\mathbf{x}'| \gg A|\mathbf{x}|$$

for some j; since $|\mathbf{x} \cdot \mathbf{b}_j| \ll |\mathbf{x}|$, the claim is again true

Returning now to the original basis $\{\mathbf{b}_0^*, \ldots, \mathbf{b}_n^*\}$, let $\ell = n$ and let j be such that $|\mathbf{x}' \cdot \mathbf{b}_j|/|\mathbf{x} \cdot \mathbf{b}_j| \gg A$, as in the claim. Then

$$|(\mathbf{x} \wedge \mathbf{x}') \cdot (\mathbf{b}_j \wedge \mathbf{b}_\ell)| = \left|\det\begin{pmatrix} \mathbf{x} \cdot \mathbf{b}_j & \mathbf{x}' \cdot \mathbf{b}_j \\ \mathbf{x} \cdot \mathbf{b}_\ell & \mathbf{x}' \cdot \mathbf{b}_\ell \end{pmatrix}\right|$$
$$\gg |\mathbf{x}' \cdot \mathbf{b}_j| |\mathbf{x}| \qquad \text{since } \mathbf{x}' \cdot \mathbf{b}_\ell = 0$$
$$\gg A|\mathbf{x} \cdot \mathbf{b}_j| |\mathbf{x}|. \qquad \qquad \square$$

PROOF OF THEOREM 6.4.3: We first prove it with a weaker bound,

$$(6.4.7) \qquad \prod_{v \in S} \frac{\|(\mathbf{x} \wedge \mathbf{x}') \cdot \mathbf{b}_i\|_v}{\|\mathbf{x}\|_v \|\mathbf{x} \cdot \mathbf{b}_i\|_v} < \overline{H}(\mathbf{x})^\epsilon.$$

Let S be the set of vectors for which (6.4.7) does not hold, and assume that the theorem is false; i. e. S is infinite. The strategy will be to apply the theory of successive minima to $V/\langle \mathbf{x} \rangle$, realized as $V \wedge \mathbf{x} \subseteq \wedge^2 V$. If no $\mathbf{x}'$ exists, then an appropriate first successive minimum is large; we then dualize and apply a few extra arguments to obtain the situation of Theorem 6.4.2.

Consider an infinite sequence of $\mathbf{x}$ for which no suitable $\mathbf{x}'$ exists. For each $0 \le i \le n$, each $v \in S$ and each $\mathbf{x}$ in the sequence, let $\mathbf{b}_{v,i,\mathbf{x}}$ be determined as in (6.4.6). Passing to an infinite subsequence, we may assume $\mathbf{b}_{v,i}$ do not depend on $\mathbf{x}$. By assumption, for all $\mathbf{x}' \in O_{k,S}^{n+1}$ with $\mathbf{x} \wedge \mathbf{x}' \ne 0$,

$$\prod_{v \in S} \frac{\|(\mathbf{x} \wedge \mathbf{x}') \cdot \mathbf{b}_{v,i}\|_v}{\|\mathbf{x}\|_v \|\mathbf{x} \cdot \mathbf{b}_{v,i}\|_v} > \overline{H}(\mathbf{x})^\epsilon$$

for some indices i depending on v and $\mathbf{x}'$. Applying the lemma, we have indices $j = j(v, \mathbf{x})$ such that,

$$(6.4.8) \qquad \prod_{v \in S} \frac{\|(\mathbf{x} \wedge \mathbf{x}') \cdot (\mathbf{b}_{v,j} \wedge \mathbf{b}_{v,n})\|_v}{\|\mathbf{x}\|_v \|\mathbf{x} \cdot \mathbf{b}_{v,j}\|_v} \gg \overline{H}(\mathbf{x})^\epsilon.$$

On $V \wedge \mathbf{x} \subseteq \wedge^2 V$, consider the system of successive minima associated to the parallelepiped given by,

$$L_{v,i}(\mathbf{x} \wedge \mathbf{x}') = (\mathbf{x} \wedge \mathbf{x}') \cdot (\mathbf{b}_{v,i} \wedge \mathbf{b}_{v,n}), \qquad 0 \le i < n;$$
$$A_{v,i} = 1/\|\mathbf{x}\|_v \|\mathbf{x} \cdot \mathbf{b}_{v,i}\|_v.$$

By (6.4.8) we then have

$$\lambda_1^{[k:\mathbb{Q}]} \gg \overline{H}(\mathbf{x})^\epsilon.$$

To compute the relative volume of $\mathbf{b}_{v,i} \wedge \mathbf{b}_{v,n}$ relative to the lattice $O_{k,S}^{n+1} \wedge \mathbf{x}$, let $x_1, \ldots, x_{n+1}$ be the coordinates of $\mathbf{x}$ relative to the standard basis $\{\mathbf{e}_i\}$, assume $x_1 \ne 0$. Then $\mathbf{e}_2 \wedge \mathbf{x}, \ldots, \mathbf{e}_{n+1} \wedge \mathbf{x}$ form a basis for a sublattice of $O_{k,S}^{n+1} \wedge \mathbf{x}$ of index $\prod_{v \in S} \|x_1\|_v$. The volume of the $\mathbf{b}_{v,i} \wedge \mathbf{b}_{v,n}$ relative to this sublattice is,

$$\prod_{v \in S} \|((\mathbf{e}_2 \wedge \mathbf{x}) \wedge \cdots \wedge (\mathbf{e}_{n+1} \wedge \mathbf{x})) \cdot ((\mathbf{b}_{v,n} \wedge \mathbf{b}_{v,0}) \wedge \cdots \wedge (\mathbf{b}_{v,n} \wedge \mathbf{b}_{v,n-1}))\|_v^{-1}$$

$$= \left[\prod_{v \in S} \|\mathbf{b}_{v,n} \wedge \mathbf{x}\|_v^{n-1} \|(\mathbf{x} \wedge \mathbf{e}_2 \wedge \cdots \wedge \mathbf{e}_{n+1}) \cdot (\mathbf{b}_{v,0} \wedge \cdots \wedge \mathbf{b}_{v,n})\|_v \right]^{-1}$$

$$\gg\ll [\overline{H}(\mathbf{x})^{n-1} \prod_{v \in S} \|x_1\|_v \| \det \mathbf{b}_{v,i}\|_v]^{-1}$$

by Lemma 6.3.13. Therefore, by Theorem 6.1.11,

$$(\lambda_1 \lambda_2 \ldots \lambda_n)^{[k:\mathbb{Q}]} \gg\ll \overline{H}(\mathbf{x})^{n-1} \prod_{v,i} A_{v,i}$$

$$\gg\ll \frac{1}{\prod_v \prod_{i=0}^n \|\mathbf{x} \cdot \mathbf{b}_{v,i}\|_v}.$$

We now apply Davenport's lemma with $\rho_1 = \rho/\lambda_i$, choosing ρ so that $\rho_1 \rho_2 \ldots \rho_n = 1$. This gives constants $\rho_{v,i}$ such that the successive minima λ_i relative to

$$f'(\mathbf{x})^{[k:\mathbb{Q}]} = \prod_{v \in S} \max_{1 \le i < n} \rho_{v,i} A_{v,i} \|(\mathbf{x} \wedge \mathbf{x}') \cdot (\mathbf{b}_{v,n} \wedge \mathbf{b}_{v,i})\|_v$$

satisfy,

$$\lambda_i' \gg\ll \prod_{v \in S} \prod_{i=0}^n \|\mathbf{x} \cdot \mathbf{b}_{v,i}\|_v^{-1/n[k:\mathbb{Q}]}.$$

From (6.2.7) we also have,

$$\rho_{v,i} \ll (\lambda_1'/\lambda_1)^{N_v} \qquad \text{for all } v \in S, \, 0 \le i < n$$

(6.4.9)
$$\ll \left(\prod_{v \in S} \prod_{i=0}^n \|\mathbf{x} \cdot \mathbf{b}_{v,i}\|_v^{-1/n} \cdot \overline{H}(\mathbf{x})^{-\epsilon} \right)^{N_v/[k:\mathbb{Q}]};$$

(6.4.10)
$$\prod_{v \in S} \max_{0 \le i < n} \rho_{v,i} \ll \prod_{v \in S} \prod_{i=0}^n \|\mathbf{x} \cdot \mathbf{b}_{v,i}\|^{-1/n} \overline{H}(\mathbf{x})^{-\epsilon}.$$

We now apply Proposition 6.3.10 to $\wedge^{n-1}(\mathbf{x} \wedge V)$. This is isomorphic to the dual of $V/\langle \mathbf{x} \rangle$, i. e. the subspace of V^* consisting of those forms vanishing at $\mathbf{x}$. By Proposition 6.3.10 and Example 6.3.3(d), the successive minima $\mu_1, \ldots, \mu_n$ satisfy,

$$\mu_i \gg\ll \prod_{v \in S} \prod_{i=0}^{n} \|\mathbf{x} \cdot \mathbf{b}_{v,i}\|_v^{-(n-1)/n[k:\mathbb{Q}]}.$$

This means that there exists a basis $\mathbf{v}_1, \ldots, \mathbf{v}_n$ of a full sublattice of the lattice $\wedge^{n-1}(\mathbf{x} \wedge O_{k,S}^{n+1})$ such that, after scaling each $\mathbf{v}_i$ by a unit, we have

(6.4.11)
$$\|\mathbf{v}_j \cdot ((\mathbf{b}_{v,n} \wedge \mathbf{b}_{v,0}) \wedge \cdots \wedge \overbrace{(\mathbf{b}_{v,n} \wedge \mathbf{b}_{v,m})} \wedge \cdots \wedge (\mathbf{b}_{v,n} \wedge \mathbf{b}_{v,n-1}))\|_v$$

$$\ll \prod_{i=0}^{n} \|\mathbf{x} \cdot \mathbf{b}_{v,i}\|^{-(n-1)/n} \cdot \prod_{\substack{i=0 \\ i \neq m}}^{n-1} 1/A_{v,i}\rho_{v,i}$$

$$\ll \prod_{i=0}^{n} \|\mathbf{x} \cdot \mathbf{b}_{v,i}\|^{-(n-1)/n} \cdot \|\mathbf{x}\|_v^{n-1} \prod_{\substack{i=0 \\ i \neq m}}^{n-1} \|\mathbf{x} \cdot \mathbf{b}_{v,i}\|_v \rho_{v,i}$$

$$\ll \frac{\|\mathbf{x}\|_v^{n-2} \rho_{v,m}}{\|\mathbf{x} \cdot \mathbf{b}_{v,m}\|_v} \prod_{i=0}^{n} \|\mathbf{x} \cdot \mathbf{b}_{v,i}\|^{1/n}$$

since $\|\mathbf{x} \cdot \mathbf{b}_{v,n}\|_v \gg\ll \|\mathbf{x}\|_v$ and $\prod_i \rho_{v,i} = 1$ for all $v \in S$. Since $\mathbf{v}_j$ lies in $\wedge^{n-1}(\mathbf{x} \wedge O_{k,S}^{n+1})$, we can write it as a finite sum,

$$\mathbf{v}_j = \sum_{\ell \in I_j} ((\mathbf{x} \wedge \mathbf{u}_{\ell 1}) \wedge \cdots \wedge (\mathbf{x} \wedge \mathbf{u}_{\ell,n-1}))$$

with $\mathbf{u}_{\ell i} \in O_{k,S}^{n+1}$. Applying Lemma 6.3.13 to the left-hand side of the above inequality,
(6.4.12)
$$\text{l.h.s.} = \|\mathbf{x}\|_v^{n-2} \left\| \sum_\ell (\mathbf{x} \wedge \mathbf{u}_{\ell 1} \wedge \cdots \wedge \mathbf{u}_{\ell,n-1}) \cdot (\mathbf{b}_{v,0} \wedge \cdots \wedge \overbrace{\mathbf{b}_{v,m}} \wedge \cdots \wedge \mathbf{b}_{v,n}) \right\|_v.$$

The vectors
$$\mathbf{u}_j = \sum_{\ell \in I_j} \mathbf{u}_{\ell 1} \wedge \cdots \wedge \mathbf{u}_{\ell,n-1}$$

lie in $\wedge^{n-1} O_{k,S}^{n+1}$, and $\mathbf{x} \wedge \mathbf{u}_j$ are linearly independent vectors spanning a full sublattice of $\{\mathbf{b} \in (O_{k,S}^{n+1})^* \mid \mathbf{x} \cdot \mathbf{b} = 0\}$. Also, noting that $\wedge^{n-1} V^* \cong V$, we write
$$\mathbf{b}_{v,m}^* = \mathbf{b}_{v,0} \wedge \cdots \wedge \overbrace{\mathbf{b}_{v,m}} \wedge \cdots \wedge \mathbf{b}_{v,n}$$

which is the dual basis to $\{\mathbf{b}_{v,0}, \ldots, \mathbf{b}_{v,n}\}$. Combining (6.4.11) and (6.4.12) then gives,

$$(6.4.13) \qquad \|(\mathbf{x} \wedge \mathbf{u}_j) \cdot \mathbf{b}_{v,m}^*\|_v \ll \frac{\rho_{v,m}}{\|\mathbf{x} \cdot \mathbf{b}_{v,m}\|_v} \prod_{i=0}^{n} \|\mathbf{x} \cdot \mathbf{b}_{v,i}\|_v^{1/n};$$

$$(6.4.14) \qquad \prod_{v \in S} \prod_{i=0}^{n-1} \max_{1 \leq j \leq n} \|(\mathbf{x} \wedge \mathbf{u}_j) \cdot \mathbf{b}_{v,i}^*\|_v \ll \prod_{v \in S} \|\mathbf{x} \cdot \mathbf{b}_{v,n}\|_v.$$

In order to apply Theorem 6.4.2, we need a bound for $\|(\mathbf{x} \wedge \mathbf{u}_j) \cdot \mathbf{b}_{v,n}^*\|_v$. But,

$$\mathbf{b}_{v,n}^* \equiv \frac{-1}{\det(\mathbf{b}_{v,i})} \sum_{i=0}^{n-1} \frac{\mathbf{x} \cdot \mathbf{b}_{v,i}}{\mathbf{x} \cdot \mathbf{b}_{v,n}} \mathbf{b}_{v,i}^* \pmod{\mathbf{x}}.$$

With (6.4.13) this gives,

$$(6.4.15)$$

$$\|(\mathbf{x} \wedge \mathbf{u}_j) \cdot \mathbf{b}_{v,n}^*\|_v \ll \frac{1}{\|\mathbf{x} \cdot \mathbf{b}_{v,n}\|_v} \prod_{i=0}^{n} \|\mathbf{x} \cdot \mathbf{b}_{v,i}\|_v^{1/n} \max_{0 \leq i < n} \rho_{v,i};$$

$$\prod_{v \in S} \prod_{m=0}^{n} \max_{1 \leq j \leq n} \|(\mathbf{x} \wedge \mathbf{u}_j) \cdot \mathbf{b}_{v,n}^*\|_v \ll \overline{H}(\mathbf{x})^{-\epsilon} \Big/ \prod_{v \in S} \|\mathbf{x} \cdot \mathbf{b}_{v,n}\|_v,$$

by (6.5.10). Combining this with (6.4.14) gives,

$$\prod_{v \in S} \prod_{m=0}^{n} \max_{1 \leq j \leq n} \|(\mathbf{x} \wedge \mathbf{u}_j) \cdot \mathbf{b}_{v,m}^*\|_v \ll \overline{H}(\mathbf{x})^{-\epsilon}.$$

Thus the second condition of Theorem 6.4.2 holds.

(The extra constant in this inequality is not a problem, since

$$c\overline{H}(\mathbf{x})^{-\epsilon} < \overline{H}(\mathbf{x})^{-2\epsilon}$$

holds whenever $\overline{H}(\mathbf{x})$ is sufficiently large. The remaining $\mathbf{x}$ are finite in number and can be included in S. For this reason in the sequel we can ignore constants implicit in $\ll$, but we will not explicitly repeat the argument each time.)

To show that the first condition also holds, we first check that if $\mathbf{x} \cdot \mathbf{b}_{v,i} \neq 0$, then

(6.4.16) $$\|\mathbf{x} \cdot \mathbf{b}_{v,i}\|_v \gg \|\mathbf{x}\|_v / \overline{H}(\mathbf{x}).$$

We obtain this by a Liouville type argument, as follows. Since $\mathbf{x}$ has integral coordinates relative to the standard basis and $\mathbf{b}_{v,i}$ is one of finitely many vectors, we have,

$$\prod_{w \in S} \|\mathbf{x} \cdot \mathbf{b}_{v,i}\|_w \gg 1;$$

also, if $w \neq v$, then

$$\|\mathbf{x} \cdot \mathbf{b}_{v,i}\|_w \ll \|\mathbf{x}\|_w.$$

Dividing these two relations gives (6.4.16).

Combining (6.4.16) with (6.4.9) gives,

$$\rho_{v,i} \ll \prod_{w \in S} \left(\frac{\overline{H}(\mathbf{x})}{\|x\|_w} \right)^{\frac{n+1}{n} \cdot N_v / [k:Q]}$$

$$= \overline{H}(\mathbf{x})^{(|S|-1)(n+1)N_v / n[k:Q]};$$

with (6.4.13) and (6.4.15) this becomes

$$\|(\mathbf{x} \wedge \mathbf{u}_j) \cdot \mathbf{b}_{v,m}^*\|_v \ll \frac{\|\mathbf{x}\|_v}{\overline{H}(\mathbf{x})} \|\mathbf{x}\|_v^{(n+1)/n} \overline{H}(\mathbf{x})^{(|S|-1)(n+1)N_v / n[k:Q]}.$$

This now implies the first condition of Theorem 6.4.2, provided that

$$\|\mathbf{x}\|_v \ll \overline{H}(\mathbf{x}).$$

It is no loss of generality to assume this, because it always holds after multiplying $\mathbf{x}$ by an appropriate unit u, and (6.4.7) is unaffected by this change.

Thus our infinite sequence of $\mathbf{x}$'s satisfies both conditions of Theorem 6.4.2, a contradiction. We have shown the existence of a vector $\mathbf{x}'$ for almost all $\mathbf{x}$, such that (6.4.7) holds. To obtain the theorem from this, multiply $\mathbf{x}'$ by a scalar unit, making all factors of (6.4.7) have the same order of magnitude. Then Theorem 6.4.3 follows (with a different ϵ).

§5. A Sketch of the Analytic Proof

As described in the introduction, let $\mathbf{x} = \mathbf{x}(t)$ be a meromorphic curve $\mathbf{x}: \mathbf{C} \to V = \mathbf{C}^{n+1}$ not lying in any hyperplane. As before, let $\mathbf{b}_0, \dots, \mathbf{b}_N$ be a set of vectors in V^* in general position. The analogue of Theorem 6.4.3 is,

THEOREM 6.5.1. *If $\mathbf{b} \in V^* \setminus \{0\}$ and $\alpha > 0$, then*

$$\alpha \int_{r_0}^r \frac{ds}{s} \int_0^s \int_0^{2\pi} \frac{|(\mathbf{x} \wedge \mathbf{x}') \cdot \mathbf{b}|^2}{|\mathbf{x}|^4} \left(\frac{|\mathbf{x}|^2}{|\mathbf{x} \cdot \mathbf{b}|^2} \right)^{1-\alpha} t \, dt \frac{d\theta}{2\pi} < 2 \, T_1(r) + c_1$$

PROOF: See [A, equation (32)]. □

To more clearly see the meaning of this integral, we have the following lemma.

LEMMA 6.5.2. *If α and β are real-valued functions such that,*

$$\int_{r_0}^r \frac{ds}{s} \int_0^s e^{\alpha(t)} t \, dt < \beta(r),$$

then

$$\alpha(r) \ll \beta(r) + 0(1). \qquad \text{//}$$

PROOF: See [A, equation (24)]. □

Thus Theorem 6.5.1 implies,

(6.5.3) $\qquad \displaystyle\int_0^{2\pi} \frac{|(\mathbf{x} \wedge \mathbf{x}') \cdot \mathbf{b}|^2}{|\mathbf{x}|^2 |\mathbf{x} \cdot \mathbf{b}|^2} \left(\frac{|\mathbf{x} \cdot \mathbf{b}|^2}{|\mathbf{x}|^2} \right)^\epsilon \frac{d\theta}{2\pi} < \epsilon \, T_1(r) + 0(1)$ $\qquad$ //

Since it holds as $\mathbf{b}$ varies over any finite collection of vectors, we see that this statement is indeed an analogue of Theorem 6.4.3. The stronger form of (6.5.1) is necessary in order to apply the averaging argument in the proof of (6.5.7), although by a longer argument it is possible to deduce (6.5.7) from (6.5.3). This result is a generalization of Theorem 6.5.1 to higher derivatives; before starting it, we introduce some notation and lemmas from [A].

Let $\mathbf{x}', \mathbf{x}'', \dots, \mathbf{x}^{(p)}$ denote the derivatives of $\mathbf{x}$; then

$$X^p = \mathbf{x} \wedge \mathbf{x}' \wedge \cdots \wedge \mathbf{x}^{(p-1)}$$

is a p-vector, which is associated to the pth osculating hyperplane. Also let the characteristic, proximity, and counting functions be defined by,

$$\overline{T}_p(r) = \left[\int_0^{2\pi} \log |X^p| \frac{d\theta}{2\pi} \right]_{t=r_0}^{t=r},$$

$$T_p(r) = 2 \int_{r_0}^r \frac{ds}{d} \int_0^s \int_0^{2\pi} \frac{|X^{p-1}|^2 |X^{p+1}|^2}{|X^p|^4} \, t \, \frac{d\theta}{2\pi} dt$$

$$= \overline{T}_p(r) - N_p(r);$$

$$\overline{T}_p(r, E^q) = \left[\int_0^{2\pi} \log |X^p \cdot E^q| \frac{d\theta}{2\pi} \right]_{r_0}^{t=r}$$

(6.5.4)
$$T_p(r, E^q) = 2 \int_{r_0}^r \frac{ds}{s} \int_0^s \int_0^{2\pi} \frac{|X^{p-1} \cdot E^q|^2 |X^{p+1} \cdot E^q|^2}{|X^p \cdot E^q|^4} \, t \, \frac{d\theta}{2\pi} dt$$

$$= \overline{T}_p(r) - N_p(r, E^q) - N_p(r);$$

$$m_p(r, E^q) = \int_0^{2\pi} \log \frac{|X^p|}{|X^p \cdot E^q|} \frac{d\theta}{2\pi};$$

$$n_p(r) = \text{no. of zeroes of } X^p \text{ inside } |z| < r;$$

$$N_p(r) = \int_{r_0}^r n_p(s) \frac{ds}{s};$$

$$n_p(r, E^q) = \text{no. of zeroes of } X^p \cdot E^q \text{ inside } |z| < r;$$

$$N_p(r, E^q) = \int_{r_0}^r n_p(s, E^q) \frac{ds}{s}.$$

For functions X^p appearing in the integrands for T_p and m_p, we mean $X^p(te^{i\theta})$. The case $p = 1$ of the above definitions correspond to the definitions of Chapter 3. The above definitions, as well as the equivalence of the expressions for T_p, can be found in [A, §3]. We also have a lemma listing some of their properties.

LEMMA 6.5.5.

(a). For every subspace E^q and every $p = 1, \ldots, n$, the following equation holds,

$$m_p(r, E^q) - m_p(r_0, E^q) + T_p(r, E^q) + N_p(r, E^q) = T_p(r).$$

(b). Fix p and a $(p-1)$-dimensional subspace E^{p-1}. Let $\mathbf{y} = (X^p \cdot E^{p-1})$; $\mathbf{y} \in V$. Then

$$T_1(\mathbf{y}, r) = T_p(\mathbf{x}, r, E^{p-1}) \le T_p(\mathbf{x}).$$

(c). *With* **y** *as above*,

$$\mathbf{y}' = (\mathbf{x} \wedge \mathbf{x}' \wedge \cdots \wedge \mathbf{x}^{(p-2)} \wedge \mathbf{x}^{(p)}) \cdot E^{p-1}$$

(d). $\overline{T}_{p-1}(r) - 2\overline{T}_p(r) + \overline{T}_{p+1}(r) < \epsilon T_p(r)$.   //
(e). $N_{p-1}(r) - 2N_p(r) + N_{p+1}(r) \geq 0$.

PROOF: (b) and (c) are trivial from the definitions, for the other parts, see [A], 3.7 and equation (21).   □

We note that (d) and (e) imply,

$$T_{p-1}(r) - 2T_p(r) + T_{p+1}(r) < \epsilon T_p(r).$$ //

Since $T_0(r) = T_{n+1}(r) = 0$, all the characteristic functions are therefore of the same order of magnitude.

Part (c) is formally very similar to the last statement in Proposition 6.3.10, since derivatives and successive minima play analogous roles.

Finally, for the proof of (6.5.7), it will be necessary to average a function of a subspace E^q, over all E^q containing a fixed E^s. To define this, first consider the case $q = s + 1$. Writing $E^q = E^s \wedge \mathbf{b}$ for some unit vector $\mathbf{b}$ orthogonal to E^s, we define the mean over E^q to be the mean over all $\mathbf{b}$ orthogonal to E^s. The general case is defined inductively. Then we have,

LEMMA 6.5.6. *For* $p \geq q \geq s$,

(a). $\underset{E^q \supseteq E^s}{\text{Mean}} \log |X^p \cdot E^q| = \log |X^p \cdot E^s| + c(p, q, s)$ *where* $c(p, q, s)$ *is a constant.*

(b). $\underset{E^{p-1}}{\text{Mean}} \log |X^p \cdot (E^{p-1} \wedge \mathbf{b})| = \log |X^p \cdot \mathbf{b}| + c(p-1, p-1, 0)$

(c). $\underset{E^{p-1}}{\text{Mean}} \, T(r, E^{p-1}) \leq T_p(r) + c(p)$.

PROOF: For (a), see [A, Lemma 1]. Part (b) follows from (a) and the fact that $(X^p \cdot (E^{p-1} \wedge \mathbf{b})) = ((X^p \cdot \mathbf{b}) \cdot E^{p-1})$. Part (c) follows from (a) and the definitions.   □

The higher dimensional version of Theorem 6.5.1 is now,

THEOREM 6.5.7. *For any nonzero* $\mathbf{b} \in V^*$ *and any* $\alpha > 0$,

$$\alpha \int_{r_0}^{r} \frac{ds}{s} \int_0^r \int_0^{2\pi} \frac{|X^{p-1}|^2 |X^{p+1} \cdot \mathbf{b}|^2}{|X^p|^2 |X^p \cdot \mathbf{b}|^2} \left(\frac{|X^p \cdot \mathbf{b}|^2}{|X^p|^2} \right)^{\alpha} t \frac{d\theta}{2\pi} dt < c \, T_p(r) + c'.$$

PROOF: We apply Theorem 6.5.1 to $\mathbf{y} = X^p \cdot E^{p-1}$. Then $\mathbf{y}'$, $\mathbf{y} \wedge \mathbf{y}'$, and $T_1(\mathbf{y})$ are given by 6.5.5c, 6.3.14, and 6.5.5b, respectively, so that,

$$\alpha \int_{r_0}^r \frac{ds}{s} \int_0^s \int_0^{2\pi} \frac{|X^{p-1} \cdot E^{p-1}|^2 |X^{p+1} \cdot (E^{p-1} \wedge b)|^2}{|X^p \cdot E^{p-1}|^4}$$

$$\times \left(\frac{|X^p \cdot E^{p-1}|^2}{|X^p \cdot (E^{p-1} \wedge b)|^2} \right)^{1-\alpha} t \frac{d\theta}{2\pi} dt$$

$$< 2 T_p(r, E^{p-1}) + c_1.$$

Take the mean over all E^{p-1}. By concavity of the logarithm,

(6.5.8)   Mean $\log < \log$ Mean.

Thus it is possible to apply Lemma 6.5.6 to the logarithm of the integrand, giving,

$$\alpha \int_{r_0}^r \frac{ds}{s} \int_0^s \int_0^{2\pi} \frac{|X^{p-1}|^2 |X^{p+1} \cdot b|^2}{|X^p|^4} \left(\frac{|X^p|^2}{|X^p \cdot b|^2} \right)^{1-\alpha} t \frac{d\theta}{2\pi} dt < 2 T_p(r) + c_p.$$

This implies the theorem.   □

In the last section, we obtained an arithmetical analogue of Theorem 6.5.1, which suffices for proving the theorem in the analytic case; yet we have not succeeded in obtaining an arithmetical version of the analytic proof. The difficulty lies in the proof of the above theorem, together with slight differences in the statements of Theorems 6.4.3 and 6.5.1. In particular, 6.5.1 is an inequality concerning the derivative $\mathbf{x}'$; whereas 6.4.3 states that vectors $\mathbf{x}'$ exist which satisfy the inequality. Thus in the arithmetic case, $\mathbf{x}'$ is not given a priori, and we do not know whether it satisfies the trivial properties satisfied by the derivative. So in attempting to translate the above proof, one could construct vectors $\mathbf{y}' \in \mathcal{O}_{k,S}^{n+1}$, but since

$$\mathbf{y} \wedge \mathbf{y}' = (X^{p-1} \cdot E^{p-1})(X^{p+1} \cdot E^{p+1}),$$

one would introduce a denominator of $X^{p-1} \cdot E^{p-1}$ in trying to find a vector $\mathbf{x}^{(p)}$ corresponding to $\mathbf{y}'$.

THEOREM 6.5.9. *For any finite collections of vectors* $\mathbf{b} \in V^*$ *in general position and any fixed* p, $1 \le p \le n$,

$$\sum_{\mathbf{b}} (m_p(\mathbf{b}, r) - m_{p+1}(\mathbf{b}, r))$$

$$\le (n + 1 - p)(-\overline{T}_{p-1}(r) + 2\overline{T}_p(r) - \overline{T}_{p+1}(r)) + \epsilon T_1(r) \quad //$$

PROOF: The integrand of 6.5.7 can be written as,

$$\frac{|X^{p-1}|^2|X^{p+1}|^2}{|X^p|^4} \cdot \Phi(\mathbf{b})$$

where

$$\Phi(\mathbf{b}) = \frac{|X^p|^{2-2\alpha}}{|X^p\cdot\mathbf{b}|^{2-2\alpha}} \Big/ \frac{|X^{p+1}|^2}{|X^{p+1}\cdot\mathbf{b}|^2}. \qquad \square$$

We now apply,

LEMMA 6.5.10 (SUMS INTO PRODUCTS). *For a fixed collection of vectors* $\mathbf{b} \in V^*$ *in general position,*

$$\sum_{\mathbf{b}} \Phi(\mathbf{b}) \geq C_{p,\mathbf{b}} \prod_{\mathbf{b}} \Phi(\mathbf{b})^{1/n+1-p}.$$

PROOF: Since X^p is decomposable and has codimension $n+1-p$, at most $n+1-p$ of the $\mathbf{b}$'s can have $|X^p\cdot\mathbf{b}|/|X^p| = 0$. Therefore there exists a constant M such that $|X^p|/|X^p\cdot\mathbf{b}| \geq M$ for at most $n+1-p$ of the $\mathbf{b}$'s. This holds true for $\Phi(\mathbf{b})$ as well. Letting $\sum'$ denote the sum over only those $\mathbf{b}$ for which $\Phi(\mathbf{b}) \geq M$, we have,

$$\log\sum \Phi(\mathbf{b}) \geq \log{\sum}' \Phi(\mathbf{b}) \geq \frac{1}{n+1-p}{\sum}' \log\Phi(\mathbf{b})$$

$$\geq \frac{1}{n+1-p}\Big(\sum \log\Phi(\mathbf{b}) - (\#\mathbf{b})\log M\Big). \qquad \square$$

Returning to the theorem, summing (6.5.7) over all $\mathbf{b}$ and applying Lemma 6.5.2 gives,

$$\log\int_0^{2\pi} \frac{|X^{p-1}|^2|X^{p+1}|^2}{|X^p|^4}\sum_{\mathbf{b}}\Phi(\mathbf{b})\frac{d\theta}{2\pi} < \epsilon\, T_p(r) + 0(1). \qquad //$$

Then Lemma 6.5.10 and concavity of the logarithm give,

$$\int_0^{2\pi}\left[\log\frac{|X^{p-1}|^2|X^{p+1}|^2}{|X^p|^4} + \frac{1}{n+1-p}\sum_{\mathbf{b}}\log\Phi(\mathbf{b})\right]\frac{d\theta}{2\pi} < \epsilon\, T_p(r) + 0(1);$$

$$//$$

$$\overline{T}_{p-1}(r) - 2\overline{T}_p(r) + \overline{T}_{p+1}(r)$$
$$+ \frac{1}{n+1-p} \sum_{\mathbf{b}} ((1 - \alpha)m_p(\mathbf{b}, r) - m_{p+1}(\mathbf{b}, r))$$
$$< \epsilon T_p(r) + 0(1). \qquad //$$

In the above discussion, α can be taken as a function of r provided it does not vanish faster than polynomially in $T_p(r)$. In particular, $\alpha(r) = 1/T_p(r)$ suffices. But by Lemma 6.5.5(a),

$$m_p(\mathbf{b}, r) < T_p(r) + 0(1);$$

therefore, $\alpha\, m_p(b, r)$ is bounded and we have the theorem, except for the difference between $T_p(r)$ and $T_1(r)$. But by the remark following Lemma 6.5.5, these are of the same order of magnitude, and we are done.

To conclude the proof of Theorem 6.0.1, sum 6.5.9 from $p = 1$ to n. Since $m_{n+1}(\mathbf{b}, r) = 0$ and $\overline{T}_{n+1}(r) = N_{n+1}(r)$, this becomes,

$$\sum_{\mathbf{b}} m_1(\mathbf{b}, r) < (n + 1 + \epsilon)T_1(r) - N_{n+1}(r) \qquad //$$

(with a different ϵ). Since $N_{n+1}(r) \geq 0$, 6.0.1(a) follows.

REMARK 6.5.11. Let us briefly consider the one dimensional case, $n = 1$. We may assume that the coordinates x_1, x_2 of $\mathbf{x}$ have no common zeroes. If x_1, say, has a zero of multiplicity e, then $\mathbf{x}'$ has a zero of multiplicity $e - 1$, so that $\mathbf{x} \wedge \mathbf{x}'$ has a zero of multiplicity $e - 1$, which then shows up in $N_2(r)$. Thus meromorphic curves $\mathbf{x}(t)$ for which (6.0.1) is close to equality will tend not to have many multiple zeroes in their coordinate functions. In the number field case, this translates into a suggestion that a similar result should hold for square divisors of coordinates of vectors for which (6.0.1) is nearly an equality. For example, coordinates of solutions of Pell's equation should be largely square free. This has also been discussed in Remark 5.4.3.

REMARK 6.5.12. If $f: \mathbf{C} \to \mathbf{P}^n$ is ramified to order e at t, then the contribution to $N_{n+1}(r)$ is at least

$$\frac{e(e+1)}{2} \log \frac{r}{|z|}, \qquad \text{if } e \leq n;$$
$$\left[en - \frac{n(r-1)}{2} \right] \log \frac{r}{|z|}, \qquad \text{if } e \geq n.$$

This gives more support for the necessity of the factor $\dim V$ in part (b) of the General Conjecture.

§6. The Remainder of the Algebraic Proof

In this section we use Theorem 6.4.3 to prove Theorem 6.0.1(b). As already noted, it is not currently possible to use a proof similar to the analytic proof of the previous section, because of minor differences between the statements of Theorem 6.4.3 and its analytic counterpart. Instead, we reduce Theorem 6.4.3 to a statement concerning successive minima, and then finish the proof by Schmidt's original method. This restatement is the following.

THEOREM 6.6.1. *Let* $\mathbf{b}_0, \ldots, \mathbf{b}_N$ *be vectors in* V^* *in general position and let* $\epsilon > 0$. *Then there exists a finite set* $S \subseteq V$ *with the following property. Assume* $\mathbf{b}_{v,0}, \ldots, \mathbf{b}_{v,n}$ *are distinct elements of* $\{\mathbf{b}_0, \ldots, \mathbf{b}_n\}$ *for each* $v \in S$ *and assume* $A_{v,i}$ *are positive real constants as in Section 2. Let* $\lambda_1, \ldots, \lambda_{n+1}$ *be the successive minima relative to the length function,*

$$(6.6.2) \qquad f(\mathbf{x})^{[k:Q]} = \prod_{v \in S} \max_{0 \le i \le n} A_{v,i} \|\mathbf{x} \cdot \mathbf{b}_{v,i}\|_v.$$

Then either $f(\mathbf{x}) = \lambda_1$ *for some* $\mathbf{x} \in kS \cap O_{k,S}^{n+1}$, *or* $\lambda_2 < \lambda_1 H(\mathbf{x})^\epsilon$.

PROOF: Let S be as in Theorem 6.4.3. Let $\mathbf{b}_{v,i}$ and $A_{v,i}$ be as in the statement of the above theorem and let $\mathbf{x} \in O_{k,S}^{n+1}$ be such that $f(\mathbf{x}) = \lambda_1$. Assume that $\mathbf{x} \notin kS$. Then by Theorem 6.4.3 there exists a vector $\mathbf{x}'$ such that,

$$(6.6.3) \qquad \sum_{v \in S} \frac{\|(\mathbf{x} \wedge \mathbf{x}') \cdot \mathbf{b}_i\|_v}{\|\mathbf{x}\|_v \|\mathbf{x} \cdot \mathbf{b}_i\|_v} < H(\mathbf{x})^\epsilon,$$

if $\mathbf{x} \cdot \mathbf{b}_i \ne 0$, and $\mathbf{x}' \cdot \mathbf{b}_i = 0$ if $\mathbf{x} \cdot \mathbf{b}_i = 0$.

Here $\mathbf{x}'$ is only determined modulo $\mathbf{x}$; let us choose one $\mathbf{x}' \in O_{k,S}^{n+1}$. For each $v \in S$, permute the vectors $\mathbf{b}_{v,i}$ such that

$$(6.6.4) \qquad \|\mathbf{x} \cdot \mathbf{b}_{v,0}\|_v \le \cdots \le \|\mathbf{x} \cdot \mathbf{b}_{v,n}\|_v.$$

It is an elementary fact of number theory that there exists a constant c_1, depending only on k, such that given constants $a_v \in k_v$ for each $v \in S$, there exists $a \in O_{k,S}$ such that

$$\|a - a_v\|_v < c_1$$

for each $v \in S$. Applying this fact with $a_v = (\mathbf{x}' \cdot \mathbf{b}_{v,n})/(\mathbf{x} \cdot \mathbf{b}_{v,n})$, we replace x' with $x' - ax$, so that

$$(6.6.5) \qquad \|\mathbf{x}' \cdot \mathbf{b}_{v,n}\|_v < c_1 \|\mathbf{x} \cdot \mathbf{b}_{v,n}\|_v.$$

But

$$\left\| \det \begin{pmatrix} \mathbf{x} \cdot \mathbf{b}_i & \mathbf{x} \cdot \mathbf{b}_{v,n} \\ \mathbf{x}' \cdot \mathbf{b}_i & \mathbf{x}' \cdot \mathbf{b}_{v,n} \end{pmatrix} \right\|_v = \| (\mathbf{x} \wedge \mathbf{x}') \cdot (\mathbf{b}_i \wedge \mathbf{b}_{v,n}) \|_v$$

$$\ll \| (\mathbf{x} \wedge \mathbf{x}') \cdot \mathbf{b}_i \|_v$$

$$\ll H(\mathbf{x})^\epsilon \| \mathbf{x} \|_v \| \mathbf{x} \cdot \mathbf{b}_i \|_v \qquad \text{(by 6.6.3)}$$

$$\ll H(\mathbf{x})^\epsilon \| \mathbf{x} \cdot \mathbf{b}_{v,n} \|_v \| \mathbf{x} \cdot \mathbf{b}_i \|_v \qquad \text{(by 6.6.4)}$$

Hence by (6.6.5),

$$\| \mathbf{x}' \cdot \mathbf{b}_i \|_v \ll (H(\mathbf{x})^\epsilon + c_1) \| \mathbf{x} \cdot \mathbf{b}_i \|_v;$$

$$\ll H(\mathbf{x})^\epsilon \| \mathbf{x} \cdot \mathbf{b}_i \|_v$$

unless $H(\mathbf{x})$ is small. Enlarging S to eliminate this possibility, we have

$$f(\mathbf{x}') \ll H(\mathbf{x})^\epsilon f(\mathbf{x}).$$

Thus,

$$\lambda_2 \ll H(\mathbf{x})^\epsilon \lambda_1,$$

(with a different ϵ), as was to be shown. $\qquad\qquad\square$

We now use this theorem to obtain a generalization of itself. This generalization corresponds roughly to Theorem 6.5.9. In both cases, we use the methods of the Grassmann algebra to reduce the general case to the case $p = 1$, but both the theorems and their respective proofs are sufficiently different that one cannot combine them and call it a common proof. On the other hand, the ideas in each case are very similar.

THEOREM 6.6.6. Let $\mathbf{b}_0, \dots, \mathbf{b}_N$ be vectors in V^* in general position, let $\epsilon > 0$, and let $1 \le p \le n$. Then there exists a finite set S_p of p-dimensional linear subspaces of V with the following property. Let $\mathbf{b}_{v,i}$, $A_{v,i}$, and $\lambda_1, \dots, \lambda_{n+1}$ be as in Theorem 6.6.1. Assume $\mathbf{x}^{(1)}, \dots, \mathbf{x}^{(n+1)}$ are linearly independent vectors in $O_{k,S}^{n+1}$ satisfying $f(\mathbf{x}^{(i)}) = \lambda_i$. Then either

$$\langle \mathbf{x}^{(1)}, \dots, \mathbf{x}^{(p)} \rangle \in S_p,$$

or

$$\lambda_{p+1} < H(\mathbf{x}^{(1)} \wedge \cdots \wedge \mathbf{x}^{(p)})^\epsilon \lambda_p.$$

PROOF: For p-tuples σ of distinct integers in the range $0 \le i \le n$, define $\mathbf{b}_{v,\sigma}$ and $A_{v,\sigma}$ as in (6.3.8). Let $\mu_1, \dots$ denote the successive minima of the associated length function on $\wedge^p V$. By Proposition 6.3.10 and Theorem 6.6.1,

$$\frac{\lambda_{p+1}}{\lambda_p} \ll \frac{\mu_2}{\mu_1} < H(\mathbf{x}^{(1)} \wedge \cdots \wedge \mathbf{x}^{(p)})^\epsilon,$$

unless $\mathbf{x}^{(1)} \wedge \cdots \wedge \mathbf{x}^{(p)}$ is a scalar multiple of some element of the finite set $S \subseteq \wedge^p V$. Let S_p be the collection of subspaces of V corresponding to decomposable elements of S. Then the conditions in the theorem hold.

$\square$

We will be applying this theorem to the successive minimum problem defined by the length function (6.6.2), where $b_{v,i}$ are chosen such that

(6.6.7)
$$\|\mathbf{x} \cdot \mathbf{b}_{v,0}\|_v \leq \cdots \leq \|\mathbf{x} \cdot \mathbf{b}_{v,N}\|_v$$

and

(6.6.8)
$$A_{v,i} = A_v / \|\mathbf{x} \cdot \mathbf{b}_{v,i}\|_v,$$

where A_v is chosen such that,

$$\prod_{i=0}^{n} A_{v,i} = 1$$

for all $v \in S$.

LEMMA 6.6.9. *Let $\lambda_1, \ldots, \lambda_{n+1}$ be the successive minima of the length function defined by (6.6.2), (6.6.7), and (6.6.8). Let $\mathbf{x}^{(1)}, \ldots, \mathbf{x}^{(n+1)}$ be linearly independent lattice points with $f(\mathbf{x}^{(i)}) = \lambda_i$. Then there exists constants c_p, $1 \leq p \leq n$, such that*

$$h(\mathbf{x}^{(1)} \wedge \cdots \wedge \mathbf{x}^{(p)}) < c_p h(\mathbf{x}) + O(1)$$

where the constants c_p depend only on k, n, and p and the constant in $O(1)$ depends only on k, n, p, and the $\mathbf{b}_i$.

(This lemma is the analogue of the remark following Lemma 6.5.5.)

PROOF: We have,

$$h(\mathbf{x}^{(1)} \wedge \cdots \wedge \mathbf{x}^{(p)}) \leq \sum_{i=1}^{p} h(\mathbf{x}^{(i)}) + O(1)$$

$$\leq \sum_{i=1}^{p} \log \lambda_1 + \frac{p}{[k:Q]} \log \max_{0 \leq j \leq n} A_{v,j} + O(1).$$

By (6.1.11), the first term is negative; it then suffices to show that,

$$\sum_{v \in S} \log \max A_{v,j} < c'_p h(\mathbf{x}) + O(1).$$

But this follows from the definition of $A_{v,j}$ and the Liouville estimate (6.4.16),

$$\frac{\|\mathbf{x}\|_v}{H(\mathbf{x})} \ll \|\mathbf{x} \cdot \mathbf{b}_i\|_v \ll \|\mathbf{x}\|_v. \qquad \square$$

We can now conclude the proof of Theorem 6.0.1 with a telescoping argument similar to the proof in the analytic case. We only need to be careful of the fact that $\mathbf{x}$ is not necessarily the first successive minimum of the parallelepiped defined as in (6.6.7) and (6.6.8).

We start with a sequence of vectors $\mathbf{x}$, and immediately throw out all $\mathbf{x}$ such that $\mathbf{x} \cdot \mathbf{b}_i = 0$ for any i, or such that $\mathbf{x}$ lies in one of the subspaces in one of the S_p. This eliminates only finitely many hyperplanes; we will show that all remaining vectors satisfy the theorem, i. e.

$$\sum_{i=0}^{N} m(\mathbf{b}_i, \mathbf{x}) \le (n + 1 + \epsilon)h(\mathbf{x}) + O(1).$$

For each vector $\mathbf{x}$, let $\mathbf{b}_{v,i}$, $A_{v,i}$, λ_1 and $\mathbf{x}^{(i)}$ be defined as in (6.6.7)–(6.6.9). Let p_0 be the smallest integer for which,

$$\mathbf{x} \in \langle \mathbf{x}^{(1)}, \ldots, \mathbf{x}^{(p)} \rangle.$$

Then Theorem 6.6.6 holds for all $p \ge p_0$.

But by (6.6.7),

$$\sum_{i=0}^{N} m(\mathbf{b}_i, \mathbf{x}) = \frac{1}{[k : \mathbb{Q}]} \sum_{v \in S} \sum_{i=0}^{N} -\log \frac{\|\mathbf{x} \cdot \mathbf{b}_{v,i}\|_v}{\|\mathbf{x}\|_v} + O(1);$$

also,

$$\log \lambda_{p_0} \le \log f(\mathbf{x})$$

$$= \frac{1}{[k : \mathbb{Q}]} \sum_{v \in S} \log A_v$$

$$= \frac{1}{(n+1)[k : \mathbb{Q}]} \sum_{v \in S} \sum_{i=0}^{n} \log \|\mathbf{x} \cdot \mathbf{b}_{v,i}\|_v;$$

$$= h(\mathbf{x}) - \frac{1}{n+1} \sum_{i=0}^{N} m(\mathbf{b}_i, \mathbf{x}) + O(1).$$

Applying (6.1.11), we also have,

$$\log \lambda_{n+1} \geq \frac{1}{n+1} \sum_{i=0}^{N} m(\mathbf{b}_i, \mathbf{x}) - h(\mathbf{x}) + O(1).$$

Thus, summing the results of Theorem 6.6.6 from $p = p_0$ to n gives,

(6.6.10)
$$\sum_{i=0}^{N} m(\mathbf{b}_i, \mathbf{x}) - (n+1)h(\mathbf{x}) \leq \frac{n+1}{2} \log \frac{\lambda_{n+1}}{\lambda_{p_0}}$$

$$\leq \epsilon \left(\frac{n+1}{2} \right) \sum_{p=p_0}^{N} \log H(\mathbf{x}^{(1)} \wedge \cdots \wedge \mathbf{x}^{(p)})$$

$$\leq \epsilon \left(\frac{n+1}{2} [k : \mathbb{Q}] \sum_{p=p_0}^{n} c_p \right) h(\mathbf{x}) + O(1)$$

by Lemma 6.6.9. This concludes the proof (after adjusting ϵ). □

§7. Conclusion

Thus we see how the proofs of Ahlfors and Schmidt, while not exactly analogous to one another, are roughly similar. This is in contrast to Griffiths' proof of the same statement for equidimensional holomorphic maps. In that case the exceptional set appears only in the assumption that the Jacobian does not vanish, whereas here it plays a direct role in the proof. It follows that, for comparison with number theory, the study of holomorphic maps on $\mathbb{C}$ should be more relevant than the study of equidimensional holomorphic maps. In particular, the Conjecture using $(1,1)$ forms is probably more tractable than the Main Conjecture, although for applications the latter is easier to use.

Also, we note that, in some sense, taking successive minima is roughly analogous to taking derivatives: they play similar roles in the two proofs given here. Moreover, the lemma of the logarithmic derivative,

$$\int_0^{2\pi} \log^+ \left| \frac{f'}{f} \right| \frac{d\theta}{2\pi} < \epsilon T(r) + O(1), \qquad //$$

implies that if $\mathbf{x} \cdot \mathbf{b}$ is small, then $\mathbf{x}' \cdot \mathbf{b}$ is likely to be small, too; therefore if $\mathbf{x}$ is the first minimum, then $\mathbf{x}'$ is a good candidate for the second minimum.

This analogy also holds, although weakly, in the function field case, where both derivatives and successive minima exist.

LEMMA 6.7.1. *Let* $k = \mathbf{C}(t)$, $S = \{\infty\}$, *and let* $L_1, \ldots, L_n$ *be* n *linearly independent linear forms with coefficients in* $\mathbf{C}$. *Let* $\mathbf{x} \in k^n$ *be a vector such that relative to the length function*

$$f(\mathbf{y}) = \max_{1 \le i \le n} |L_i(\mathbf{y})|_\infty / |L_i(\mathbf{x})|_\infty$$

the first successive minimum is 1. *If* $\lambda_2 > 1$ *then*

$$\lambda_2 = f(\mathbf{x}').$$

PROOF: Trivial.

The proof breaks down in the non-split case.

BIBLIOGRAPHY

[A]   Ahlfors, L. V., The theory of meromorphic curves, Acta Soc. Sci. Fenn. N. S. A Tom III (1941) pp. 1–31.

[Ar 1]   Arakelov, S. Ju., Families of algebraic curves with fixed degeneracies, Izv. Akad. Nauk SSSR Ser. Mat. 35 (1971) pp. 1277–1302.

[Ar 2]   —————— , Intersection theory of divisors on an arithmetic surface, Math. USSR. Izvestija 8 (1974) pp. 1167–1180.

[Ar 3]   —————— , Theory of intersection on the arithmetic surface, Proc. Intl. Congress of Mathematicians, ed. by Ralph D. James, Canadian Mathematical Congress, Vancouver, 1975, Vol. 1, pp. 405–408.

[Bo]   Borel, E., Sur les zèros des fonctions entières, Acta Math. 20 (1896) pp. 357–396.

[B-V]   Bombieri, E.; Vaaler, J., On Siegel's lemma, Inv. Math. 73 (1983) 11–32; Addendum, Inv. Math. 75 (1984), p. 377.

[C-G]   Carlson, James; Griffiths, Phillip, A defect relation for equidimensional holomorphic mappings between algebraic varieties, Ann. Math. 95 (1972) pp. 557–584.

[Car 1] Cartan, H., Sur les systèmes de fonctions holomorphes à variétés linéaires lacunaires (Thése) Ann. Sci. de l'Ecole Normale Superieure 45 (1928) pp. 255–346.

[Car 2] —————— , Sur les zéros des combinaisons linéaires de p fonctions holomorphes données, Mathematica 7 (1933) pp. 5–31.

[Cas]   Cassels, J. W. S., An introduction to the geometry of numbers, Grundlehren der Mathematischen Wissenschaften 99, Springer-Verlag, Berlin-Göttingen-Heidelberg, 1959.

[Co-G]   Cowen, M.; Griffiths, P. A., Holomorphic curves and metrics of negative curvature, J. d'Analyse Math. 29 (1979) pp. 93–153.

[Dan]   Danilov, L. V., Diophantine equation $x^3 - y^2 = k$ and Hall's conjecture, Mat. Zametki 32 (1982) 273–275; English translation, Math. Notes of the USSR 32 (1982) pp. 617–618.

[**Dav**]   Davenport, Harold, On $f^3(t) - g^2(t)$, <u>K</u>. <u>Norske</u> <u>Vid</u>. <u>Selskabs</u> <u>Farh</u> (Trondheim), 38 (1965) pp. 86–87.

[**F**]   Faltings, G., Endlichkeitssätze für abelsche Varietäten über Zahl-körpern, <u>Inv</u>. <u>Math</u>. 73 (1983) pp. 349–366.

[**Fr 1**]   Frey, G., Links between stable elliptic curves and certain dio-phantine equations, <u>Annales</u> <u>Universitatis</u> <u>Saraviensis</u>, <u>Series</u> <u>Mathematicae</u> 1 (1986) pp. 1–40.

[**Fr 2**]   ———— , Letter to Serge Lang, 3 Sept. 1986.

[**Gr 1**]   Green, Mark L., Some Picard theorems for holomorphic maps to algebraic varieties, <u>Amer</u>. <u>J</u>. <u>of</u> <u>Math</u>., 97 (1975) pp. 43–75.

[**Gr 2**]   ———————— , Some examples and counterexamples in value distribution theory for several variables, <u>Compositio</u> <u>Math</u>., 30 (1975) pp. 317–322.

[**Greub**]   Greub, W. H., <u>Multilinear</u> <u>Algebra</u>, Grundlehren der mathema-tischen Wissenschaften 136, Springer-Verlag, New York, 1967.

[**G**]   Griffiths, P. A., <u>Entire</u> <u>holomorphic</u> <u>mappings</u> <u>in</u> <u>one</u> <u>and</u> <u>several</u> <u>variables</u>: Hermann Weyl Lectures, The Institute for Advanced Study, Princeton University Press, Princeton, NJ, 1976.

[**G-K**]   Griffiths, P. A.; King, James, Nevanlinna theory and holomor-phic mappings between algebraic varieties, <u>Acta</u> <u>Math</u>. 130 (1973) pp. 146–220.

[**Hal**]   Hall, Jr., Marshall, The diophantine equation $x^3 - y^2 = k$, in <u>Computers</u> <u>in</u> <u>Number</u> <u>Theory</u>, ed. by A. O. L. Atkin and B. J. Birch, Academic Press, London, 1971, pp. 173–198.

[**H**]   Hartshorne, R., <u>Algebraic</u> <u>Geometry</u>, (Graduate Texts in Mathe-matics, 52) Springer-Verlag, New York-Heidelberg-Berlin, 1977.

[**Ii**]   Iitaka, S., <u>Algebraic</u> <u>Geometry</u> (Graduate Texts in Mathematics, 76), Springer-Verlag, New York-Heidelberg-Berlin, 1982.

[**K**]   Kobayashi, S., <u>Hyperbolic</u> <u>Manifolds</u> <u>and</u> <u>Holomorphic</u> <u>Map-pings</u>, Marcel Dekker, Inc., New York, 1970.

[**K-O**]   Kobayashi, S.; Ochiai, T., Mappings into compact complex manifolds with negative first Chern class. <u>J</u>. <u>Math</u>. <u>Soc</u>. <u>Japan</u> 23 (1971), pp. 137–148.

[L 1]   Lang, S., Some theorems and conjectures in diophantine equations, <u>Bull</u>. <u>AMS</u>, 66 (1960) pp. 240–249.

[L 2]   ———, Integral points on curves, <u>Publ</u>. <u>Math</u>. <u>IHES</u> 6 (1960) pp. 27–43.

[L 3]   ———, <u>Introduction</u> <u>to</u> <u>Diophantine</u> <u>Approximations</u>, Addison-Wesley, Reading, Mass., 1966, p. 71.

[L 4]   ———, <u>Algebraic</u> <u>Number</u> <u>Theory</u>, Addison-Wesley, Reading, Mass., 1970.

[L 5]   ———, Higher dimensional diophantine problems, <u>Bull</u>. <u>AMS</u>, 80 (1974) pp. 779–787.

[L 6]   ———, <u>Elliptic</u> <u>Curves</u>: <u>Diophantine</u> <u>Analysis</u>, Grundlehren der Mathematischen Wissenschaften 231, Springer-Verlag, Berlin, 1978, p. 213.

[L 7]   ———, <u>Fundamentals</u> <u>of</u> <u>Diophantine</u> <u>Geometry</u>, Springer-Verlag, New York, 1983.

[L 8]   ———, Hyperbolic and diophantine analysis, <u>Bull</u>. <u>AMS</u>, 14 (1986) pp. 159–205.

[L 9]   ———, <u>Introduction</u> <u>to</u> <u>Complex</u> <u>Hyperbolic</u> <u>Spaces</u>, Springer-Verlag, New York, 1986.

[L 10]   ———, Conjectured diophantine estimates on elliptic curves, in <u>Arithmetic</u> <u>and</u> <u>Geometry</u>, M. Artin and J. Tate, eds., Birkhäuser, Boston, 1983; pp. 155–172.

[L 11]   ———, Diophantine approximations on toruses, <u>Amer</u>. <u>J</u>. <u>of</u> <u>Math</u>. 86 (1964) pp. 521–533.

[Lau]   Laurent, M., Equations diophantiennes exponentielles, <u>Inv</u>. <u>Math</u>. 78 (1984) pp. 299–327.

[Mac]   MacFeat, R. B., Geometry of numbers in adele spaces, <u>Dissertationes</u> <u>Mathematicae</u> (<u>Rozprawy</u> <u>Matematyczne</u>) LXXXVIII, Warszawa 1971.

[Mah]   Mahler, K., An analogue to Minkowski's geometry of numbers in a field of series, <u>Ann</u>. <u>Math</u>. 42 (1941) pp. 488–522.

[Man]   Manin, Yu. I., Lectures on the K-functor in algebraic geometry, Russ. Math. Surveys 24 (1969), pp. 1-89.

[Mas 1]   Mason, R.C., Diophantine Equations over Function Fields, (London Mathematical Society Lecture Note Series, 96), Cambridge University Press, Cambridge, 1984.

[Mas 2]   _____ , Equations over function fields, in Number Theory, proceedings of the Noordwijkerhout 1983, (Lecture Notes in Mathematics, 1068), Springer-Verlag, Berlin-Heidelberg-New York (1984) pp. 149–157.

[Mat]   Matsusaka, T., The criterion for algebraic equivalence and the torsion group, Amer. J. of Math. 79 (1957) pp. 53–66.

[Maz]   Mazur, B., Arithmetic on Curves, Bull. AMS 14 (1986) pp. 207–259.

[Min]   Minkowski, H., Diophantische approximationen, Leipzig, B.G. Teubner, 1907, Chapter 6.

[M-M]   Mori, S.; Mukai, S., The uniruledness of the moduli space of curves of genus 11, Algebraic Geometry Conference, Tokyo-Kyoto, 1982, (Lecture Notes in Mathematics; 1016), Springer-Verlag, Berlin-Heidelberg-New York, 1983, pp. 334–353.

[Mum]   Mumford, D., Algebraic Geometry I (Grundelhren der mathematischen Wissenschaften, 221), Springer-Verlag, Berlin-Heidelberg-New York, 1976.

[N]   Noguchi, J., A higher dimensional analogue of Mordell's conjecture over function fields, Math. Ann. 258 (1981) pp. 207–212.

[Na]   Nagata, M., Imbedding of an abstract variety in a complete variety, J. Math. Kyoto Univ. 2 (1962) pp. 1–10.

[Os 1]   Osgood, C. F., A number theoretic-differential equations approach to generalizing Nevanlinna theory, Indian J. of Math. 23 (1981) pp. 1–15.

[Os 2]   _____ , Sometimes effective Thue-Seigel-Roth-Schmidt-Nevanlinna bounds, or better, J. Number Theory 21 (1985) pp. 347–389.

[**Ray**]   Raynaud, M., Hauteurs et isogenies, in Séminaire sur les pinceaux arithmetiques; la conjecture de Mordell, Astérisque 127, (1985) pp. 199–234.

[**Rey**]   Reyssat, E., Une remarque sur des theorems de Picard et Siegel, in Approximation diophantienne de fonctions de Weierstrass, thesis, Paris VI, 1982.

[**R-SD**]   Rogers, K.; Swinnerton-Dyer, H. P. F., The geometry of numbers over algebraic number fields, Trans. AMS 88 (1958) pp. 227–242.

[**R**]   Roth, K. F., Rational approximations to algebraic numbers, Mathematika 2 (1955) pp. 1–20.

[**Schl 1**]   Schlickewei, H. P., Linearformen mit algebraischen koeffizienten, Manuscripta Math. 18 (1976) pp. 147–185.

[**Schl 2**]   ——————— , On products of special linear forms with algebraic coefficients, Acta Arith. 31 (1976) pp. 389–398.

[**Schl 3**]   ——————— , The $\wp$-adic Thue-Siegel-Roth-Schmidt theorem, Archiv der Math. 29 (1977) pp. 267–270.

[**Schl 4**]   ——————— , Über die diophantische Gleichung $X_1 + X_2 + \ldots + X_n = 0$, Acta Arith. 33 (1977) pp. 183–185.

[**Schm 1**]   Schmidt, Wolfgang M., Diophantine Approximation, (Lecture Notes in Mathematics, 785); Springer-Verlag, Berlin-Heidelberg-New York, 1980.

[**Schm 2**]   ——————— , Norm form equations, Ann. Math. 96 (1972) pp. 526–551.

[**Sha**]   Shafarevich, I. R. *et. al.*, Algebraic surfaces, Proceedings of the Steklov Institute of Mathematics, Amer. Math. Soc., Providence, RI, 1967, Vol. 75.

[**Sil 1**]   Silverman, J. H., The Catalan equation over function fields, unpublished. See also The Catalan equation over function fields, Trans. AMS 273 (1982) pp. 201–205.

[**Sil 2**]   ——————— , The S-unit equation over function fields, Math. Proc. Camb. Phil. Soc. 95 (1984) pp. 3–4.

[**Sil 3**]   ——————— , Integral points on abelian varieties, preprint, 1984.

[Sil 4]   —————— , Arithmetic distance functions and height functions in diophantine geometry, preprint.

[St]   Stoll, W., Value distribution on parabolic spaces, (Lecture Notes in Mathematics, 600), Springer-Verlag, Berlin, 1973.

[Sz 1]   Szpiro, L., Propriétés numériques du faisceau dualisant relatif, in Séminaire sur les pinceaux de courbes de genre au moins deux, Astérisque 86 (1971), pp. 44–78.

[Sz 2]   —————— , La conjecture de Mordell, Exposé Bourbaki 619, Astérisque 121–122 (1985), pp. 83–103.

[Tot]   Totaro, B., Proof of a conjecture of Lang, preprint, Aug. 1986.

[vdP]   van der Poorten, A. J., The growth conditions for recurrence sequences, unpublished, July, 1982.

[Vo 1]   Vojta, P., Integral points on varieties, thesis, Harvard, 1983.

[Vo 2]   —————— , A higher dimensional Mordell conjecture, Arithmetic Geometry, ed. by Cornell, G. and Silverman, J. H., Graduate Texts in Mathematics, Springer-Verlag, New York, 1986, pp. 341–353.

[W-W]   Weyl, H.; Weyl, J., Meromorphic curves, Ann. Math. 39 (1938) pp. 516–538.

[W]   Wirsing, E., On approximation of algebraic numbers by algebraic numbers of bounded degree. Proc. of Symp. in Pure Math. XX (1969 Number Theory Institute) (1971) pp. 213–247.

Index

abc conjecture, 71–72, 84
Abelian varieties, 23, 35, 47–48, 82–84
Absolute value, 1
  equivalent, 1
Ahlfors, L. V., 89
Algebraic integral point, see Integral
  points
Algebraic point, 77
Almost ample, 4, 42, 43
Ample line bundle, 41
Arakelov, S. Ju., 39, 40
Arithmetic distance function, 24–25,
  28
Associated cycle, 25
Asymptotic Fermat conjecture, 71, 84,
  88

Blowing-up, 13, 25, 26, 45, 48, 52, 61,
  70, 75, 76
Bombieri, E., 46, 94–95
Borel lemma, 24

Carlson, J., 42, 50
Cartan, E., 89
Characteristic function, 32, 33, 34, 80,
  113
Chern class, 41
Chevalley-Weil theorem, 14, 57–61, 64
Concavity of the logarithm, 81, 115
Conductor, 85–87
Conjecture using (1,1) forms, 51, 80,
  122
Counting function, 31, 33, 34, 40, 77,
  113
Cowen, 89
Currents, 80–81
Curvature, 48–50, 80

d(*k*), see Discriminant term
Danilov, L. V., 51
Davenport's lemma, 96–98, 108
Decomposable vector, 99
Defect, 30, 32, 34, 34, 36–38, 42
Degenerate, 15, 22, 39, 46, 48

$\delta(D)$, see Defect
Diagonal hyperplane, 19, 27, 28–29
Different, 58
Differentials, 47, 48, 58
Dimension, of divisor, 4
Dirichlet unit theorem, 18, 27
Discriminant term, 58, 64, 69, 78
Div *V*, 21

Elliptic curves, 74, 86–87
Exceptional Zariski-closed subset, 43,
  56, 67, 69, 76
Exceptional subspace, 27–29

Faltings, G., 82
  See also Shafarevich conjecture,
    Mordell conjecture
Fermat, see Asymptotic Fermat conjec-
  ture
First Main Theorem, 32, 34, 36, 41
Frey, G., 84, 87
Fubini-Study metric, 41
Function fields, 69, 77–79, 122–123

Gaussian curvature, 49
General Conjecture, 63, 66, 68–70,
    70–76, 78, 84, 117
General position, 19, 25, 90
General type, 4, 46
Generalized diagonal, 28–29
Generically finite, 58
Genus formula, 74, 79
Green, M. L., 16, 20, 24, 44
Griffiths function, 48–49
Griffiths, P. A., 39, 42, 50, 89

h(), see Height
H(), see Height
Hadamard, 31
Hall's conjecture, 51–56, 72–74
Hall-Lang-Stark conjecture, 74
Hall-Lang-Waldschmidt-Szpiro conjec-
  ture, 84, 85–87, 88
Height, 2, 8, 33, 34, 65, 77, 96
  absolute, 3

Height (*continued*)
  equivalent, 3
  logarithmic, 3
  multiplicative, 3
  quasi-equivalent, 5
  relative, 3
Hermite-Minkowski theorem, 14
Holomorphic map, 15, 24, 30, 80
Hurwitz, see Genus formula
Hyperbolicity, 48–50

Iitaka, S., 47–48
Integral point, 10–15, 44, 47, 47–48, 83
  algebraic, 12
Integralizable set of points, 11
Interior product, 102
Intersection pairing, 12, 77

Jensen's formula, 34, 35

K-theory, 62
K3 surface, 23
Kobayashi, S., 48–50
Kodaira Embedding Theorem, 41
Kodaira, 4
Kwack, 50–51

λ_v, see Weil function
Lang, S., 20, 23, 35, 47–48, 74
  see also Hall-Lang-Stark conjecture
Laplace identity, 100, 101
Large (1,1) form, 50–51
Laurent, M., 19
Length function, 92, 102, 120
Length, 58
Liouville argument, 111
Liouville theorem, 36
log^+, 31
Logarithmic canonical bundle, 47
Logarithmic general type, 47, 48

$m(a,r)$, see Proximity function
M_k, 2
Mahler, K., 94
Main Conjecture, 42, 46, 48, 54, 62,
  70, 122
Mason, R. C., 51, 74, 79, 85
Masser, D., 71, 84
Matsusaka, T., 6
Mean (over subspaces), 114
Measure hyperbolic, 49–50

Metrized line bundle, 7, 9, 39, 41
Minkowski, 90
Moduli space, 82–84
Monodromy, 15
Mordell conjecture, 46, 47
Mordell-Weil theorem, 21

$N(a,r)$, see Counting function
$N_1(r)$, see Ramification term
Nagata, M., 12
Nakai-Moishezon criterion, 6
Negatively curved, 49
Néron-Severi group, 16, 21
Noguchi, J., 46
Non-degenerate, see Degenerate
Norm form equation, 37
Normal crossings divisor, 42, 44, 62
Northcott, 6
Numerically equivalent divisor, 6

Oesterlé, J., 71, 84
Osgood, C., 30, 79

Pell's equation, 70, 117
π_1, 15
Picard group, 21
Picard variety, 16
Positive line bundle, 41
Product formula, 2, 34, 35
Proximity function, 31, 33, 34, 39, 40,
  77, 113
Pseudofication, 49–50

Ramification divisor, 40, 58, 68
Ramification point, 80
Ramification term, 32, 61–64, 80–83
Raynaud, M., 84
Reyssat, E., 30
Ric(), 49
Roth's theorem, 16–17, 18, 30, 34, 36,
  43–44, 65, 70
  see also Schmidt's theorem

S_∞, 2
Scheme, 12
Schlickewei, H. P., 16, 17, 19
Schmidt's theorem, 18, 26, 38, 43–44
Schmidt, W. M., 16, 17, 37, 65–66, 89
Second Main Theorem, 32, 34, 42
Shafarevich conjecture, 51, 80
Shafarevich, I. R., 23

Siegel's theorem, 30, 37, 47
Silverman, J. H., 47, 51, 74, 79
  see also Arithmetic distance function
Size, 17, 18
Stark, H., see Hall-Lang-Stark conjecture
Stoll, W., 39, 42
Successive minima, 90–96, 107, 114, 120, 122, 123
Sums into Products, 116
Szpiro, L., 64, 84, 86–87

$T(r)$, see Characteristic function
Totaro, B., 50

Units, 19–21

Vaaler, J., 94–95
Van der Poorten, A. J., 16, 19

Waldschmidt, M., see Hall-Lang-Waldschmidt-Szpiro conjecture
Weil function, 7–9, 19, 25–26, 28, 52, 77
  global, 7
  local, 7
Weil height, see Height
Weyl, H. and J., 32, 89
Wirsing, E., 65

Zariski topology, 15, 22, 39, 42, 43